D1256499

ADVANCE PRAISE FOR *THE STORY OF CO₂*

"Ozin and Ghoussoub take us on a captivating journey through the good, the bad, and the ugly of a quintessential molecule of our universe: CO_2. The book provides a uniquely holistic view on CO_2, spanning its multiple dimensions – scientific, technological, socioeconomic, political – and connects them into a comprehensive story. It will leave a lasting footprint on anyone susceptible to one of the greatest challenges of humankind: the renewable energy conundrum."

Bettina V. Lotsch, Nanochemistry Department, Max Planck Institute for Solid State Research

"Ozin and Ghoussoub provide an interesting narrative of a small molecule with an outsized impact. Spanning the atomic to the applied, this book is easily accessible, making it appropriate for those who want to attain foundational knowledge key to understanding pressing energy and environmental issues and their effect on society."

Chad A. Mirkin, Director of the International Institute for Nanotechnology, Northwestern University

"*The Story of CO₂* addresses possibly the most important issue facing the global community: anthropogenic CO_2 and the consequent interrelated issues of energy, global and regional economies, and the ways in which we live and work. It presents the science in a clear and comprehensible way. It is not free from controversy and will stimulate argument and debate, but the story of CO_2 is one that needs to be told and to which we all need to listen."

Richard Catlow, Department of Chemistry, University College London

"This is a comprehensive yet easily understandable book that teaches about one of the most challenging problems human beings face – carbon dioxide – and the possible solutions. I highly recommend this excellent book."

Yi Cui, Materials Science and Engineering, Stanford University

"*The Story of CO₂* addresses the big picture of the role of CO_2 in modern industry. It discusses cutting-edge scientific research on the greenhouse effect and its relation to climate change, with detailed discussion about the control of CO_2 emissions and sustainable CO_2 conversion."

Tierui Zhang, Key Laboratory of Photochemical Conversion and Optoelectronic Materials, Technical Institute of Physics and Chemistry (TIPC), Chinese Academy of Sciences (CAS)

THE STORY OF CO$_2$

BIG IDEAS FOR A SMALL MOLECULE

GEOFFREY A. OZIN AND
MIREILLE F. GHOUSSOUB

ÆVO UTP

Aevo UTP
An imprint of University of Toronto Press
Toronto Buffalo London
utorontopress.com
© University of Toronto Press 2020

This book is presented solely for educational purposes and is designed to provide helpful information on the subjects discussed. Though effort has been made to ensure accuracy of the information contained therein at the time the authors finished writing the book in October 2019, the authors and publisher do not warrant the contents of this manuscript to be free of unintentional errors or omissions.

Products and firms, institutions, agencies, and public offices mentioned are strictly illustrative examples. Being mentioned does not imply an explicit or implicit recommendation nor does it imply an explicit or implicit warranty. Likewise, the fact that a competing product, firm, institution, agency, or public office is not mentioned does not imply an explicit or implicit criticism nor does it imply an explicit or implicit concern.

Library and Archives Canada Cataloguing in Publication

Title: The story of CO$_2$: big ideas for a small molecule / Geoffrey A. Ozin and Mireille F. Ghoussoub.
Names: Ozin, Geoffrey A., author. | Ghoussoub, Mireille F., 1991– author.
Description: Includes bibliographical references and index.
Identifiers: Canadiana (print) 20200285556 | Canadiana (ebook) 20200285750 | ISBN 9781487506360 (cloth) | ISBN 9781487533960 (EPUB) | ISBN 9781487533953 (PDF)
Subjects: LCSH: Carbon dioxide. | LCSH: Carbon dioxide mitigation – Technological innovations.
Classification: LCC QD181.C1 O95 2020 | DDC 551.51/12 – dc23

ISBN 978-1-4875-0636-0 (cloth)
ISBN 978-1-4875-3396-0 (EPUB)
ISBN 978-1-4875-3395-3 (PDF)

Printed in Canada

We acknowledge the financial support of the Government of Canada, the Canada Council for the Arts, and the Ontario Arts Council, an agency of the Government of Ontario, for our publishing activities.

Canada Council **Conseil des Arts**
for the Arts **du Canada**

ONTARIO ARTS COUNCIL
CONSEIL DES ARTS DE L'ONTARIO
an Ontario government agency
un organisme du gouvernement de l'Ontario

Funded by the Financé par le
Government   gouvernement
of Canada   du Canada

To our partners, Linda and Rafa

Contents

Preface

Nothing in life is to be feared, it is only to be understood. Now is the time to understand more, so that we may fear less.

– Marie Skłodowska Curie

Academic research scientists tend to have a reputation for being socially and politically disengaged. Somehow, even in 2019, the myth of the solitary, disheveled professor locked up in an office, drowning among stacks of dusty books, and scribbling incoherent notes on pads of paper still pervades our notion of academic research. Those of us at universities know that this stereotype could not be further from the truth and that strong collaboration and effective communication are fundamental to the success of the modern research scientist. What is true, however, is that the obligations of both professors and graduate students in the present day are certainly demanding: grant writing, teaching, research, and administrative duties can easily fill the better part of the week. Wandering away from one's focused research program is not trivial and often not condoned within the community.

The research, writing, and consulting that culminated in this text certainly proved to be an exercise in wandering. We carry scientific expertise in nanochemistry and materials science. Our research involves studying the catalytic processes that might enable carbon dioxide (CO_2) to react chemically and form other compounds. However, like everyone else, we are living in the time of the existential threat of climate change. Halting the release of carbon dioxide into the atmosphere has become imperative if we are to cap global temperature rise to maintain a livable environment on Earth. Naturally, we became curious as to how our research on carbon dioxide fit into the big picture of a global climate-change-mitigation strategy. As it turns out, although our lab benches are a far cry from the scale of industrial chemical plants, fundamental research in nanochemistry and materials science is an essential part of improving on current carbon-related technologies and industrial processes.

Since the release in the 1960s of the first cautionary reports predicting global warming, climate change has slowly disseminated its way to the public consciousness and become what is arguably the biggest issue of our time. While opinions will always vary, many people, irrespective of political leaning or economic conviction, recognize that climate change is associated with anthropogenic carbon emissions and that there is the need for definitive action. At the time of writing this study, in September 2019, no less than seven million people took part in a global climate strike. With an ever-growing youth-led movement and an increasing number of countries and municipalities declaring a state of climate emergency, the urgency of the matter could not have been more evident. The Green New Deal, conceived in the spirit of Franklin D. Roosevelt's New Deal in response to the Great Depression, to enable a just and equitable transition toward an emissions-free economy is now a proposed legislation in the United States, and other countries are quickly following suit in developing their own version. At present, the social and political forces surrounding climate change are shifting daily, and we recognize that, by the time of reading, some of our content will already be outdated.

While the urgency to mitigate emissions has finally become apparent in mainstream media, the feasibility of breaking away entirely from our current fossil fuel–based economy remains continuously called into question. Electric vehicles provide a clear substitute for combustion-fuel-engine cars, but no such tangible solution seems to yet exist for air travel. From fertilizer in our fields to rubber in our bicycle tires, most chemicals and materials used today are still derived from fossil products. The ideas presented herein aim to make the case that it is technologically possible for all our modern industries – from power generation to transportation to agriculture to manufacturing – to operate without releasing further carbon dioxide into the atmosphere. Renewably generated electricity, hydrogen fuel, and carbon dioxide itself all have an important role to play in the transition toward an emissions-free economy. At the heart of this story, however, is the notion that carbon dioxide can be not only captured and sequestered but also *transformed* into a myriad of products, from feedstock chemicals to polymers to cement to drugs to synthetic fuels. Through these pages we attempt to shine light on the latest scientific research and technologies that will hopefully leave you with a clearer picture of what a future emissions-free world might look like.

Before proceeding any further, we should emphasize that while science can explain the origins and effects of climate change, it alone cannot resolve the problem. Climate change is a complex environmental, economic, and social issue that requires comprehensive, systems-level solutions. While science and technology inevitably play an important role in reducing our emissions, solutions are only as effective as their implementation. As we shall see, we are already too advanced to expect that good engineering and market signaling alone can eradicate our carbon emissions. Informed policy, strong public support, grassroots leadership, and unwavering political will are just as, if not more, crucial than technological know-how in achieving the systemic changes required to transition our planet toward a just, equitable, and low-carbon future.

So, although we try our best to outline strategies, we cannot offer a comprehensive plan to address climate change. This said, much has already been written on the topic, and we refer the reader to a list of recommended reading in appendix B. Instead, this story seeks to inspire a scientifically inclined audience as well as the general-interest reader on the latest carbon dioxide technologies and how they fit within the broader context of carbon-emission-mitigation strategies.

The Story of CO_2: Big Ideas for a Small Molecule is a seven-chapter account of all aspects of carbon dioxide, from the atomic to the universal. Rather than assuming a strictly technical viewpoint, the story interweaves fundamental scientific concepts with economics, engineering, and technology. Each chapter is followed by a bullet-point summary of the key concepts to take away. Chapter 1, "The Good, the Bad, and the Oily," provides an overview of the world's changing energy economy, highlighting our ongoing dependence on fossil-based energy sources. The chapter starts by introducing the Paris Agreement, then explores in greater depth the current global energy mix and the projected global energy trends. The state of both our primary energy sources and our trends in energy consumption are discussed. The chapter concludes that, despite the increasing market share of renewable-energy suppliers and technologies, our growing population's heavy dependency on fossil fuels means that these numbers are forecasted to grow in the immediate future.

Chapter 2, "From Space to Earth and Back Again," begins with the CO_2 molecule's origins in the big bang and its distribution throughout the planetary atmospheres of the solar system. We then address the role of CO_2 in the creation of life on Earth, the process of biomineralization, and the earth's natural carbon cycle. The exchange of carbon between the four main carbon reservoirs is described in greater detail, supplemented by a discussion on the use of carbon-isotope-tracing technology for tracking the source of carbon on our planet. We then offer a detailed explanation of the greenhouse effect and walk the reader through a simple mathematical model from which one can predict the expected rise in temperature given a small change in atmospheric conditions. The chapter concludes by discussing the effects of rising levels of atmospheric CO_2 on our climate and our health.

In contrast to the first two chapters, chapter 3, "Confronting Climate Change," addresses the social, political, and ethical aspects of addressing climate change. It begins by identifying four attitudes that pose a barrier to action on climate change and goes into further detail on how income, lifestyle, and identity help to shape our perspective on the environment and climate. It briefly covers the topic of climate justice and the need to work toward solutions that are effective, equitable, and just.

Chapter 4, "Stubborn Emissions," details the reasons that certain industrial processes are more challenging to decarbonize than others. It introduces the concept of carbon capture and storage and explains how capturing, recycling, and repurposing CO_2 can help to enable a faster energy transition.

Chapter 5, "Power to the CO_2," covers the technical challenges of recycling CO_2 and introduces the various ways in which energy, in the form of heat, electricity, or sunlight, can help to convert it into useful products. It goes on to address the other key aspects of CO_2 technologies – namely, hydrogen, water, and renewable electricity – and how they might be sourced in a sustainable manner.

Chapter 6, "It Is a CO_2 World," uses the information presented in the previous chapters to illustrate more concrete examples of the CO_2 capture, storage, and utilization technologies that might be integrated into our industries. It covers everything from fertilizer production to concrete manufacturing to air-conditioning.

Finally, chapter 7, "Bringing It All Together," illustrates the big-picture requirements of bringing these technologies to market and shares a vision of what CO_2 refineries might look like in the future. It also provides a list summary of the key chemicals and materials needed to sustain the manufacturing of critical commodities, and how they can be obtained without fossil resources. The chapter concludes by offering a practical list of ways in which the individual reader can make a difference.

Provided in appendix A is an exhaustive (though not complete) list of existing companies whose products and processes are based on CO_2 utilization technologies. Appendix B recommends further reading. By the end of this book, the reader should have a solid understanding of how carbon-dioxide-conversion technologies can offer significant benefits for the economy, environment, and society.

Short vignettes are sprinkled throughout the text as optional asides for the more curious reader who wishes to delve deeper into certain topics. Given the highly interdisciplinary spirit of this book, all technical terms are defined in the margins to assist readers with new jargon. The text includes schematics and illustrations to enrich the reader's experience and to make the text as engaging as possible, especially to non-experts. We have provided citations for those who wish to explore further any point or to refer to original sources; the references presented at the end of the book are numbered sequentially as they first appear in the text. Still, we recognize that some of the material covered in these chapters may be very new to readers with little or no background in science. Most of the technical detail can be overlooked without compromising the key message of each chapter. So, although we encourage the readers to challenge themselves in learning new material, we fully endorse a selective reading of the book to guarantee a joyful and satisfactory experience for individuals of all educational backgrounds.

As scientists, we are excited to share some of the latest emerging science that is transforming the capacity for carbon dioxide to be recycled. We believe this story to be critical to today's current political and cultural climate, in which the public receives conflicting messages about effective solutions to combating climate change. We also think that scientists owe a responsibility to the wider community to inform it of the potential impact of new technologies. Evidence-based decision-making is our only viable guide to navigate effectively the changing energy landscape, and the key instigator of political action on carbon mitigation and climate change will ultimately be an informed public.

In the spirit of Marie Curie's statement, we hope that *The Story of CO$_2$* will inspire in you a newfound appreciation of and optimism for the non-fossil CO$_2$ molecule and all it can offer.

Acknowledgments

I'm very conscious of the fact that you can't do it alone. It's teamwork. When you do it alone you run the risk that when you are no longer there nobody else will do it.
— Wangari Maathai, *The Green Belt Movement: Sharing the Approach and the Experience*

The Story of CO_2 would not have been possible without the help and support of numerous colleagues and friends.

First and foremost, we thank both past and present members of the Solar Fuels team at the University of Toronto: Paul Duchesne, Meikun Xia, Thomas Wood, Alexandra Tavasoli, Young Li, Nhat Truong Nguyen, Jon Babi, Wendy He, Abhinav Mohan, Thomas Dingle, Abdinoor Jelle, Camilo Viasus, Zhao Li, Lourdes Hurtado, Yang Fan Xu, Yuchan Dong, Navid Soheilnia, Athan Tountas, Wei Sun, Chenxi Qian, Leo Diehl, Annabelle Wong, Joller Wang, Kulbir Kaur Ghuman, Jia Jia, Laura Reyes, Paul O'Brien, and Laura Hoch. Your research efforts, creative energy, and unfailing teamwork have served as a tremendous motivator throughout the writing process. Most importantly, our thanks go to Sue Mamiche-Afara and Tamika Clarence for ensuring that the wheels of the group kept turning no matter what. We are forever grateful to be able to work with such an incredibly talented team of diverse and hard-working individuals.

An enormous thank-you goes to Leah Connor, Stephen Jones, and the rest of the team at the University of Toronto Press for believing in the story and helping us navigate the publishing process.

Our thanks go to colleagues both at home and around the world who have lent their brilliant minds to discussing everything from surface chemistry to microreactor design to solar thermal engineering to climate economics. To Peter Styring, Mohini Sain, Roland Dittmeyer, Aldo Steinfeld, Christos Maravelias, Chandra Veer Singh, Hermenegildo Garcia, Thomas Mallouk, Charles Mims, Robert Morris, Dvira Segal, Ben Hatton, Keith Butler, and Aron Walsh: your collaboration, support, creativity, and technical advice over the years have been invaluable to many of the ideas expressed in this book.

A huge thank-you goes to Erik Haites, environmental economist at Margaree Consultants; Jeffrey MacIntosh, professor in the Faculty of Law, University of Toronto; Richard Hall of Desjardins Insurance; and Robert Davies, lawyer, whose appraisals and editing of the book's contents have proved to be invaluable.

Mireille thanks colleagues Alexandra Tavasoli, Molly Sung, Ellen Gute, and Stafford Sheehan, as well as Hannah Ellix, Melanie Vasselin, Fauziya Issa, and Aleksandar Arsovski, for their invaluable conversation and feedback. She also thanks Joseph, Michelle, Louise, and Nassif for their unwavering support and encouragement.

Finally, we thank our partners, Linda and Rafa, for their love and patience throughout the tumultuous journey that is the writing process. We could not have done it without you.

Ode to CO$_2$

O small molecule,

A friend or a foe,

To love or to hate,

To understand you better,

Before it is too late,

And we all become CO$_2$.

– Geoffrey A. Ozin, *Advanced Science News*, May 9, 2012

1

The Good, the Bad, and the Oily

If there is one thing that we in the global north know how to do, it is to *consume*. The product-packed shelves that line the aisles of stores and the seemingly infinite options brought to us through on-line shopping can leave us feeling overwhelmed by choice. A *Los Angeles Times* report suggested that the average American house-hold contains roughly 300,000 items.[1] While we do not encourage anyone to start counting them, there is no denying that even those living "minimalist" lifestyles in wealthier countries are likely to own thousands of products, from clothes to electronics to furniture to medicine.

In general, the flow of goods in our current economic model can be described as a linear process: products are manufactured (often from multiple other chemicals and products), packaged, distrib-uted, purchased, consumed, and ultimately disposed. Introducing recycling to the process can, at least to some extent, transform the linear economy into one that is circular; that is, instead of the prod-uct ending up in a landfill, some materials might be recovered and reused to begin the cycle anew.

The consumer economy viewed from the perspective of a chem-ist or a chemical engineer, however, might look a little different. Rather than conceive the flow of raw materials through to consumer products as linear or circular, a scientist or engineer might imagine a tree, as shown in figure 1. In this analogy the roots represent the

Figure 1. **The chemical tree** in which raw materials, such as minerals, wood, water, and fossil resources, are located at the roots, and the consumer products that they form make up the tree's leaves.

most basic raw materials available to us, such as plants, minerals, water, and air. Moving up the trunk to the crown, we reach the tree's largest branches. These correspond to the set of basic products that can be made from the raw materials. Smaller branches shoot off from these larger ones, and even smaller branches from those, eventually yielding twigs with leaves. The point is that the most basic raw materials, when taken in combination, can form a larger set of products, which in turn can be used to create an even larger set. The process continues to the point where we are able to create all the complex products and consumer goods available today. The tree analogy especially makes sense when one considers that as many as 100,000 chemicals and consumer products derive from a mere few hundred chemicals and that these intermediate chemicals are made from roughly twenty basic chemicals, which in turn come from the base natural resources – gas, coal, oil, minerals, water, and air.

Unlike consumer goods such as cars and clothes, most of the materials and chemicals located in the pathway between the root raw materials and the final consumer products are mainly invisible to the average consumer. Taking the example of seemingly simple adhesive tape, if we were to work our way *down* the tree of chemicals, starting from the roll of tape, to the root resources, we would easily encounter several dozen chemicals and materials along the way. The adhesive used to make the sticky side of the tape is composed of an acrylate polymer, which itself derives from a type of acrylate monomer, which in turn is made by reacting acrylic acid with an alcohol. Acrylic acid is made using propylene, a by-product of gasoline production, meaning that both the acid and the alcohol originate from crude oil. The non-sticky backing of the tape is made from cellulose acetate, which itself comes from acetic acid and cellulose, the latter of which is obtained from the fibers in wood or cotton. To allow the tape to be wound and unwound, the backing is often treated with a release coating, such as fluorosilicone, a synthetic rubber made from petroleum products. Finally, the backing and adhesive are stuck together using a styrene acrylic or a polyurethane, both of which can be made from a wide range of chemicals; most of these chemicals start from oil-derived products such as propylene, ethylene, and ammonia. All this, however, does not even include the chemicals, materials, and water that enable the many manufacturing processes, or all those that go into making the plastic dispenser and packaging.

Clearly there is more to a simple everyday roll of adhesive tape than meets the consumer's eye.

You might have been surprised to find that crude oil – a fossil fuel – is a raw material used in the making of adhesive tape. For many of us, putting fuel in our cars might appear as our most direct contact with the fossil industry; however, the truth is that we live in a veritable fossil-fuel economy in which the majority of consumer products – for example, aspirin, rubber, paint, plastics, and fertilizer – derive from chemicals that themselves derive from fossil products. The point is that if we are to break free completely from our reliance on fossil resources, without compromising the manufacturing of essential consumer goods, we need to find alternative ways of making that set of basic chemicals located near the bottom of the chemical tree.

Do substitutes for these chemical processes even exist? The answer is surprising. As it turns out, there is plenty of potential for CO$_2$, arguably an abundant natural carbon resource, to replace fossil products in the manufacturing of many key chemicals. Although you may be skeptical, we encourage you to keep an open but critical mind as we take you through this journey of the CO$_2$ molecule, from its creation at the beginning of the universe to its movement throughout the earth's carbon cycle to its role in inducing global warming and finally to its potential to help decarbonize our industries, reduce dependence on fossil resources, and ultimately mitigate the impacts of climate change. First, however, the following is some background into our emissions crisis.

It Is All Up in the Air

Le Bourget is a small municipality located in the northeastern suburbs of Paris. It was here, in December of 2015, that all 196 members of the United Nations Framework Convention on Climate Change agreed to adopt the first universal, legally binding, climate-change deal, better known as the Paris Agreement. It opened for signature on April 22, 2016, in honor of Earth Day, and officially came into effect in November of the same year, one month after fifty-five members, who accounted for at least 55 percent of global emissions, had ratified the agreement. The document outlines goals and strategies to strengthen the global response to climate change, the most fundamental of which is to keep global temperature rise well below 2°C, with an aspiration of 1.5°C.

Global temperature rise is directly related to the amount of CO_2 in the atmosphere, as we shall explain in greater detail in chapter 2. Much like how a doctor measures blood pressure to gauge a patient's overall health, measuring CO_2 concentrations is a key tool in monitoring the health of the planet. The National Oceanic and Atmospheric Administration (NOAA) of the United States has been recording atmospheric carbon dioxide levels over the past fifty years. These data, recorded in parts per million (ppm), have been presented in the form of the well-documented Charles Keeling Curve, shown in figure 2.

The overall trend shows a continuous increase in atmospheric CO_2 concentrations over the last fifty years, from 315 ppm in 1958 to 408 ppm in 2020. The periodically spaced black wiggles superimposed onto the curve correspond to the annual rise and fall of CO_2 concentrations in accordance with seasonal plant growth and decay.

The continuous rise in atmospheric CO_2 levels has been connected to the use of fossil fuels since the beginning of the industrial revolution. Frighteningly, over half of the total amount of greenhouse gases emitted since the end of the eighteenth century occurred during the past thirty years, due to the accelerated extraction and use of fossil fuels.[4] While we could easily fill pages presenting the overwhelming scientific evidence of the rapid, human-induced global warming, many excellent books have already been written on the topic.

Figure 2. **Keeling Curve, showing atmospheric carbon dioxide, in parts per million, from the years 1700 to 2020.**

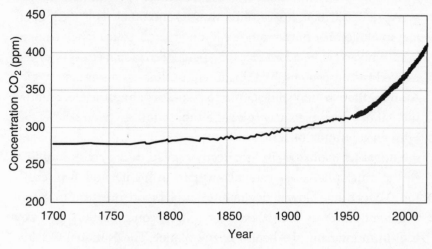

Courtesy of Scripps Institution of Oceanography.[2,3]

Every year the World Meteorological Organization (WMO) releases a report on the status of the global climate. To date it has declared the past five years (2015, 2016, 2017, 2018, and 2019) to be the warmest on record,[*] with 2016 being the warmest of them all at 1.1°C above pre-industrial times.[5] This has been accompanied by consistently rising levels of atmospheric CO$_2$, dramatic drops in Arctic sea ice, and a major rise in global sea levels each year. In 2018 the global temperature of the ocean reached the highest on record, contributing to coral bleaching and mortality in tropical waters.[6] Rising temperatures are also having detrimental effects on glaciers, with up to one-third of the Himalayan range, which currently serves as a water source to over a billion and a half people, expected to melt by 2100.[7] Multiple catastrophic weather events include record-breaking

---

[*]   Although 2018 was the coolest of the five years, it was marked by the La Niña phenomenon, which characteristically begets lower temperatures and therefore should not be misinterpreted as a reversal of the rising temperature trend.

hurricanes in the Caribbean Sea and the Atlantic Ocean, over forty million people displaced by flooding in the Indian subcontinent, severe drought in East Africa, and bush fires in Australia.[5,8,9] Although the term *climate refugee* is not yet recognized under international law, the extreme weather and increased frequency of natural disasters caused by climate change is forcibly displacing millions of people from their communities. Conflict resulting from climate-related food-and-water insecurity further exacerbates the human toll of climate change. Although it is difficult to estimate the exact numbers, the Internal Displacement Monitoring Centre reported 18.8 million new disaster-related internal displacements in 2017, with floods and storms being the biggest trigger.[10]

According to the *Global Energy and CO$_2$ Status Report 2019* of the International Energy Agency (IEA), annual global energy-related CO$_2$ emissions grew by 1.7 percent in 2018, reaching a historic high of 33.1 gigatonnes (Gt)[*] – that is over 33 billion tonnes in a single year.[11] To help us grasp this reality, figure 3 shows a picture of a ten-meter-diameter sphere of CO$_2$, weighing one tonne, in relation to a London double-decker bus. This illustration gives meaning to one tonne of CO$_2$ in the form of volume under ambient temperature and pressure conditions.[†]

Imagine 33 billion of these one-tonne CO$_2$ spheres being injected into our atmosphere every year from the combustion of fossil fuel. This is equivalent to putting one thousand of these CO$_2$-filled balloons into the atmosphere every *second*. To evade action on climate change is no longer a choice but a failure to adapt to reality, with serious environmental, social, and economic implications.

The two main strategies in the Paris Agreement to combat climate change involve mitigating greenhouse gas (GHG) emissions

---

[*] The total number of greenhouse gas emissions is over 50 Gt a year if we include the contribution of methane, nitrous oxide, and fluorinated gases.[12]

[†] Specifically, this corresponds to the volume of CO$_2$ at 1 atm (the atmospheric pressure at sea level), at which the density of CO$_2$ gas is 1.98 kg/m^3.

Figure 3. **Volume** equivalent of one tonne of CO_2 in relation to the size of a London double-decker bus.

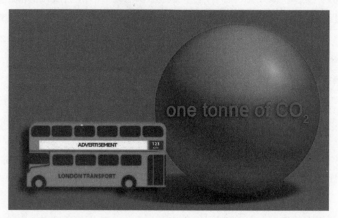

Illustration courtesy of Dr. Chenxi Qian.

and reducing fossil fuel consumption. So how are we doing when it comes to curbing our emissions to meet our goals?

Each year for the past decade an international team of leading scientists has put forth the *UN Environment Emissions Gap Report* to assess the world's progress in meeting the Paris Agreement goals of holding global temperature to well below 2°C. It evaluates the level of implementation of member countries' emission-reductions pledges with regard to current emission trends to create an "interim report card" of our efforts so far.

The findings of these reports have been discomforting. The national pledges at the foundation of the Paris Agreement are expected to cover only one-third of the emissions reductions necessary to remain on track toward the goal of staying well below 2°C by 2100.[13] At this point, if all countries follow through completely with both their conditional and their unconditional emission-reduction promises, we could still face a 3°C rise in global temperature by the end of the century. Countries will need to go above and beyond their initial pledges if we are to have a chance at keeping warming well below 2°C. To make matters worse, many countries are already failing to meet their originally proposed pledges for 2020. New policies and swifter action will be necessary for many member states, especially given that the ambiguity of some of the proposed nationally determined contributions caused uncertainty in the emissions originally estimated for 2030.[14]

The New Climate Economy: The 2018 Report of the Global Commission on the Economy and Climate warns that the next ten to fifteen years present a critical "use it or lose it" opportunity, emphasizing in particular that the next two to three years constitute a critical window.[15] At this point there is considerable uncertainty as to whether the energy transition from fossil fuels to renewables, and the concomitant reduction in CO_2 emissions, will be fast enough, especially given the current fossil fuel outlook and lagging political action.[16] In November 2018 the WMO announced that global temperatures were on course for a rise of 3°C to 5°C this century, far overshooting the global target of limiting their increase to 2°C. This is particularly alarming given that economic transitions have historically occurred over the span of decades.[17] Clearly we need to act quickly to get the world on track to meet the emission targets. But what are these targets exactly, how do they relate to global temperature rise, and where does the challenge lie in reducing them?

As previously mentioned, global temperature rise is directly related to the amount of CO_2 in the atmosphere. Scientists can therefore estimate the "carbon budget" that ensures a decent chance at limiting global temperature rise to a given level. *Global Warming of 1.5°C*, the 2018 report of the Intergovernmental Panel on Climate Change (IPCC), states that keeping our cumulative CO_2 emissions to no more than 420 Gt (that is, 420×10^9 tonnes) of CO_2 will result in a 66 percent chance of global temperature rising to no more than 1.5°C above pre-industrial levels.

What exactly do we mean by *cumulative emissions*? These are the total CO_2 emissions that are added to the atmosphere over a certain time. As we shall see in chapter 2, this number is actually different than the amount we send into the atmosphere, because nearly half of emissions are taken up by the ocean and terrestrial biospheres. The IPCC's report considers cumulative emissions as those starting in 2018. At present we emit roughly 33 Gt of CO_2 into the atmosphere annually, and this number continues to grow every year. From 2010 to 2014, for example, emissions increased at an unprecedented annual rate of 2.75 percent. So, to cap our *total* cumulative emissions at 420 Gt, it is not enough to simply reduce our annual emissions to a steady rate; we need to emit fewer and fewer emissions every year until we eventually reach *zero* annual emissions. Recent climate reports from the IPCC, the National Academies of Sciences, Engineering, and Medicine, and the Royal Society, all state that ceasing emissions entirely might not be enough and that removing CO_2 from the atmosphere will be necessary to prevent a dangerous rise in global temperature.

How fast exactly do these reductions need to happen? If we overshoot our target, will we still be able to reduce global temperatures? The 2018 IPCC report suggests that CO_2 emissions should peak to no more than 30 Gt per year by 2030, then start to decline, ideally reaching zero emissions by 2050. While these recommendations are based on climate models, which are inherently subject to assumptions and uncertainties, one thing remains clear: the faster we reduce our annual emissions to zero the better chance we have of restricting global temperature rise.

Trends in Global Emissions

Before diving into the ways in which we can cut our GHG emissions, we begin by identifying the derivations of the emissions. Our planet's total GHG emissions can be broken down in many ways: by country or region, by person, by source, and so forth. One particularly useful approach is to break down emissions according to the following sectors: energy, industry, transportation, buildings, and land use.

Breaking down emissions according to each of these five sectors, however, is not simple. For example, buildings account for nearly 20 percent of all GHG emissions, but this is in large part due to buildings consuming large amounts of heat and electricity (via lighting, ventilation systems, heating, and air-conditioning, for example). In fact, over half the emissions from buildings are indirect, associated with the production of energy in the form of heat and electricity. So, although these emissions appear to come from buildings themselves, they ultimately derive from the energy sector.

From this example, one can quickly appreciate why it is important to identify the scope of these sectors. For the purpose of discussing **decarbonization** strategies, we will stick to breaking emissions down according to their sector of origin. For example, the production of energy worldwide is responsible for nearly 35 percent of all GHG emissions.* Although much of this energy is eventually used by other sectors (mainly buildings and industrial processes), these emissions result from the production of the energy itself, and therefore we will attribute them to the energy sector.

Decarbonization: the elimination of greenhouse gas emissions, specifically carbon dioxide, from a technology, process, or system.

The fact that most of our GHG emissions derive from energy production should come as no surprise given that most of our energy, be it heat or electricity, comes from burning oil, natural gas, and coal. While heat and electricity production account for over two-thirds of these emissions, the remaining third comes from other auxiliary processes in the energy supply chain, such as fuel extraction, refining, and processing.

We tend to refer to oil, natural gas, and coal together as simply *fossil fuels*. While this is accurate in that all three derive from fossil

---

* The production and consumption of energy together account for over 70 percent of all GHG emissions resources, it is worth noting that they are not equal in terms of their emissions. Coal-fired power plants were the single largest contributor to the growth in emissions in 2018, with coal-fired electricity generation accounting for nearly one-third of total CO_2 emissions. In fact, a recent assessment conducted by the IEA found burning coal to be the single largest source of global temperature rise.[18] In certain instances, switching from coal to natural gas, a less carbon-intensive fossil fuel, has the potential to reduce the energy sector's emissions. For example, switching from coal to natural gas in 2018 alone helped to avoid 40 megatonnes (Mt) and 45 Mt of CO_2 emissions in the United States and China, respectively.[18] We emphasize, however, that substituting natural gas for coal does not necessarily offer a sustainable solution to the mitigation of emissions, and the overall carbon footprint of natural gas depends on the nature of its extraction. For example, obtaining natural gas contained within shale formations requires hydraulic fracturing technology to enable a process otherwise known as fracking. This involves drilling deep underground to access the shale formation and then pumping high-pressure fluid to crack the rock, thereby releasing any gas trapped within. One study found that when methane emissions are included, the greenhouse-gas footprint of shale gas is larger than that of conventional oil, natural gas, and coal.[19] So, while it might be tempting to view natural gas as the "cleaner" fossil-based energy source, the true environmental impact of any one fossil-fuel project depends on the specifics of its extraction and refinement processes. Although coal use is decreasing worldwide, its decline is being outpaced by increased consumption of oil and natural gas, which are now the principal drivers of growth in carbon dioxide emissions.[20] Be it coal, natural gas, or oil, burning fossil fuels is the main cause of climate change.

It may come as a surprise that some of the largest contributors of emissions, after energy, are agriculture, forestry, and land use. These sectors generally consist of all human activity involving land changes. Unlike the energy sector's emissions that derive from the burning of fossil fuels, the emissions associated with land-use change often result from alterations to natural carbon sinks. For example, the removal of natural forests to expand agricultural lands contributes 6 Gt of CO_2 emissions every year. Deforestation in the tropics, for example, accounts for over 4 Gt of CO_2 emissions alone.[21] This sector also includes many of the emissions that arise from our global food-supply chain. You may have heard that going vegetarian is one of the most impactful measures you can take to reduce your individual carbon footprint. Indeed, at 7.1 Gt CO_2-equivalent per year, livestock contribute more GHG emissions than any other food source, and cattle alone account for two-thirds of these.[22]

There are many promising strategies to reduce emissions associated with agriculture, forestry, and land use. Many of these simply involve implementing more sustainable management practices, rather than completely overhauling the infrastructure. For example, integrating trees and shrubs into fields dedicated to crops and livestock, a practice known as **agroforestry**, can help prevent erosion, enhance soil properties, improve water infiltration, and ultimately increase the carbon-storage capacity of the land through photosynthesis. Similarly, other land-management strategies have the potential to ameliorate the land's carbon-sequestration capacity. Conservation, restoration, and improved land-management strategies have the potential to mitigate up to 23 Gt of emissions per year.[23]

Agroforestry: the practice of growing trees and/or other perennial plants alongside crops and/or livestock.

Technological strategies also exist to address emissions associated with agriculture, forestry, and land use. Advances in molecular biology are creating opportunities to restore and enhance the carbon uptake of land and to adapt crops to changing climates. Genome modification, for example, can help both to enhance the health and yield of crops and to make plant varieties that may thrive under warmer temperatures. Other strategies, such as vertical farming, in which crops are grown on inclined surfaces or integrated into other structures, can minimize the amount of land dedicated to agriculture. Incorporating trees and plants into urban landscapes presents another opportunity to enhance the carbon-sequestration of land. The award-winning Bosco Verticale (Vertical Forest), a pair of residential towers in Milan that boasts hundreds of trees and tens of thousands of different plants and shrubs on its facades, removes upwards of thirty tonnes of CO$_2$ every year. Greening buildings not only constitutes a feasible and effective strategy for capturing carbon but also can help regulate building temperature, promote local biodiversity, and clean the air – not to mention being an aesthetic addition to the urban landscape. For example, PhotoSynthetica, a United Kingdom–based company, provides a variety of building cladding solutions that capture CO$_2$ from air, using algae. Its technology enhances the sequestration capacity of vertical areas by maximizing the interaction of the air flow with the algae-containing cladding, resulting in the removal of approximately one kilogram of CO$_2$ per day.

Can't We Just Plant More Trees?

Biomass: plant or animal material, including wood, agricultural crops, organic residues, algae, and seaweeds. The concept of restoring the earth's natural **biomass** as a strategy to mitigate carbon emissions would seem sensible. After all, the largest flux of CO$_2$ between the atmosphere and land occurs via plant photosynthesis. This alone amounts to an incredible 440 Gt of CO$_2$ being removed from the atmosphere every year. A 2019 study published in *Science* concluded that an additional 4.4 billion hectares of forest canopy could potentially store a further 200 Gt or more of carbon.[24] However, planting more trees is not quite as simple a solution as it may seem.

Reforesting lands that historically had no forest canopy, such as grasslands and savannas, can seriously risk damaging native ecosystems by causing widespread loss of habitats and, in turn, a severe loss of biodiversity.[25] Another concern is that tree planting without careful planning can result in agriculture being conducted on land that was once occupied by native forests instead of on land that is being reforested.[26] Reforestation projects must therefore be tailored to an area's unique socio-ecological context if they are to be successful.[27]

Key to the impact of tree planting is the rate at which forests can sequester carbon, which depends on the specific plant species, soil composition, location, and climate. The complex interplay between the world's diverse forest ecosystems and the changing climate continues to generate debate within the scientific community about the exact global carbon-capture potential of trees. Biological carbon-removal efforts alone will likely not suffice to counteract emissions, because they are severely limited by land availability; in addition, reforestation does not necessarily provide a short-term solution, because it would take several decades for forests to reach the maturity required to sequester carbon.

This said, the scientific community is not divided on the importance of global biomass restoration – the protection and restoration of native ecosystems, particularly boreal and rain forests, which form the earth's natural terrestrial carbon reservoirs and can serve to ensure the long-term stabilization of the carbon cycle.

Today, there is fascinating research emerging into the creation of new plant varieties with a boosted capacity for CO_2 capture. Dr. Joanne Chory of the Salk Institute for Biological Studies and the Howard Hughes Medical Institute is cross-breeding plants so that they contain more suberin, a long-lasting carbon-storing compound that is familiar to us as cork. According to Dr. Chory, if about 5 percent of the world's farmland were growing highly enriched suberin crops, it would fix 50 percent of all CO_2 emissions.[28]

While technologies such as these might hold the key to combating climate change in the future, some critics still argue that the massive losses of food-productive land and the elimination of natural ecosystems that would incur from employing present large-scale biological fixation strategies do not justify their gains. Still, if they were sustainably managed and monitored, some degree of biomass-based solutions could certainly contribute to the global emission-reduction strategy.[29] If anything, priority should be given to slowing the rapid rate of removal of native forests around the world. Shifting the focus to stopping deforestation, rather than to planting trees, is likely more effective in increasing the number of trees on the planet.

The next most intensive sector is industry, accounting for over a fifth of global emissions. The industrial sector comprises a broad range of manufacturing processes, including those of chemicals, pharmaceuticals, steel, iron, and cement. Roughly 40 percent of industry's emissions derive just from the on-site burning of fossil fuels to produce the heat required for these processes.[30] For example, the production of cement involves heating limestone (carbon carbonate) to very high temperatures to decompose it into calcium oxide and carbon dioxide. Cement production therefore emits carbon dioxide in two ways: first through the burning of fossil fuels to generate enormous amounts of heat, and second by nature of the chemical reaction itself, which yields CO_2 as a by-product. The fact that many industrial emissions are intrinsic to the chemical processes themselves makes it particularly challenging to address. Although switching to renewable energy can eliminate the emissions associated with powering industrial processes, it cannot resolve the CO_2 produced by the reactions themselves; however, we will delve deeper into this later.

Discussions around GHG emissions often center on cars: the need to drive less, to switch to electric vehicles (EVs), and to improve transportation infrastructure. Contributing nearly 15 percent of our total GHG emissions, the transportation sector has major potential to decarbonize.* Similarly to the energy sector, most emissions from transportation arise from the combustion of fossil fuels and therefore have the potential to be replaced by renewably generated electricity. Even without the introduction of new technologies, significant reductions may be achieved through improving the efficiency of internal combustion engines, expanding public transit systems, and bettering operational efficiency in transport systems. For example, some estimate that aviation emissions could be reduced by up to two-thirds simply through the implementation of operational changes to enhance efficiency.[31] Alternative viable options for travelers, such as high-speed electric trains for short intercity trips, could also significantly reduce the number of flights taking off every day.

Finally, buildings are another major sector behind global carbon emissions. The direct emissions associated with buildings refer to the amount of emissions produced on site. If we also consider the emissions associated with the production of heat and electricity that is then consumed by buildings, the latter account for nearly one-fifth of energy-related CO_2 emissions. Clearly, there is the potential for major emission savings by simply improving building design and switching to more efficient heating and cooling systems, lighting, and appliances. The United Nations Environment Programme estimates that increasing building efficiency could prevent nearly 2 Gt of CO_2 emissions per year.[13] All the more reason to invest in energy-efficient appliances and switch to LED (light-emitting diode) bulbs – it makes sense for both your wallet and the environment.

Hopefully you now have a better idea of where all these emissions come from. That is an important first step to identifying where

---

* The greenhouse gas emissions resulting from transportation vary dramatically across the globe and are heavily tied to unequal patterns in consumption. In wealthier countries, for example, where there are more cars per capita, transportation generally accounts for a larger share of total carbon emissions.

reductions are needed most if we are to meet the goals laid out by the Paris Agreement. While emissions from all sectors must be addressed if we are to eventually bring our net annual emissions to zero, reducing the emissions associated with the energy sector is especially critical given its connection to other sectors, particularly industry, transportation, and buildings. At this point we will briefly digress to discuss global trends in energy because we believe they warrant special attention.

Global Trends in Energy

The enormous amount of emissions currently associated with the energy sector illustrates the importance of transitioning toward non-fossil energy generation. Technologies that utilize renewable energy sources, such as photovoltaic cells, wind turbines, tidal power, and geothermal heating systems, are already proving to be capable of large-scale heat and electricity production and are occupying an increasing portion of the global energy sector.

The growth in renewable energy can be better appreciated by observing the trends in global shares of primary energy since 1990, shown in figure 4. The chart displays the amount of energy produced, expressed in kilotonne oil-equivalent, from fossil-based sources (i.e., oil, natural gas, and coal) and from non-fossil energy sources, worldwide. Note that the same quantity of two different fuels can yield different amounts of energy; therefore, to make a just comparison, the amounts have been expressed in terms of kilotonne of oil equivalent, that is, the amount of energy released by burning one thousand tonnes of crude oil.*

---

* The carbon content of fossil fuels does not scale with the energy they provide because of the different organic content in each fossil fuel. One tonne of natural gas contains 0.75 carbon content and yields 50.2 gigajoules (GJ) of energy. The same amount of oil and coal contains 0.84 and 0.85 carbon content but yields 41.6 and 26.0 GJ of energy, respectively. For the same reason, the exact hydrocarbon content of a barrel of crude oil varies somewhat between locations.

Figure 4. **Global primary energy supply**.

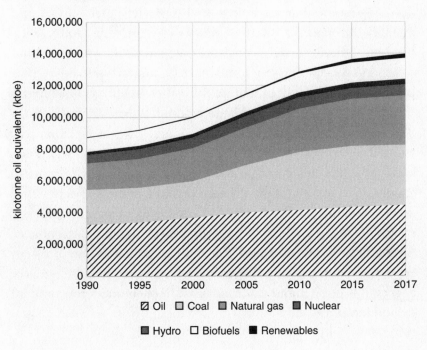

Source: IEA (2019), World Energy Balances.

Hydroelectric power (hydro) has been excluded from the category of renewable energy sources. It involves harnessing the energy of flowing water through a turbine to produce electricity via a generator. The fact that water is not consumed in the process makes hydro theoretically a renewable source of energy; however, it is not included in the category of renewable energy sources largely because of its effect on the environment. Large-scale hydro typically requires the construction of dams that have an impact on natural water systems, aquatic ecosystems, and the silt loads of rivers and streams. Dams can block migrating fish from reaching spawning grounds and can even affect the temperature and chemistry of rivers. The carbon footprint of hydroelectric power has also turned out to be higher than previously thought. The artificial reservoirs, characteristic of hydro projects, give off CO_2 and methane through the decay of previously stable soil and vegetation. Put otherwise, dam construction transforms what was once a carbon-sequestering environment (e.g., forests, wetlands) into a carbon-emitting environment. A 2016 study analyzed 1,500 hydro plants from around the world and computed that a global average of roughly 173 kilograms (kg) of CO_2 and 2 kg of methane was emitted per megawatt hour (MWh) of electricity produced from hydro.[32] While this is much lower than the emissions from fossil energy, it is still significant. The impact of hydro projects varies greatly, and many design measures can be taken to minimize their effects on aquatic wildlife. Dams can also be upgraded to enhance their efficiency and to avoid construction of brand-new projects. So, while hydro is technically a lower-carbon, renewable source of energy, it is not without environmental impacts and a concomitant carbon footprint.

Flammable Ice: The Last Fossil Fuel

You might be surprised to know that there are enormous reserves of methane (natural gas) contained within ice. Methane clathrate, also known as methane hydrate, is a compound consisting of cage-like structures of ice, called clathrate, within which lie methane gas molecules. To the naked eye, they look like regular ice crystals; however, the large quantities of methane contained within them are deemed a major source of untapped fossil fuel: a single cubic meter of methane hydrate provides around 160 cubic meters of methane gas.

Methane hydrates form when temperatures are low enough, such as those in the Arctic, and in deep water where the pressure is sufficiently high. As such, they can exist in sediments under the ocean floor, as well as beneath permafrost on land. Methane can be generated thermogenically, created by heating deep underground, or biogenically through microbial action in the top layers of sediment and soil. When clathrates melt, the previously trapped methane is released to the surrounding environment. Safe extraction of the gas is technologically challenging, energy intense, and expensive, with the added difficulty of not accidentally releasing the trapped methane into the atmosphere. China, Japan, India, and South Korea, countries devoid of resource-based oil, are actively involved in mining methane hydrates from the sea. The United States and Canada are also actively mining methane hydrates from under the permafrost in Alaska and northern Canada.

Unfortunately methane may leak during extraction, which, as well as its very use as a fuel, will only exacerbate the GHG problem. The extent of damage that its release could incur is particularly worrisome when one considers that the amount of trapped methane in the Arctic alone is around four to five times greater than the amount of CO_2 we have emitted since the time of the industrial revolution. In fact, the most dramatic rise in global temperature is not expected to come directly from increasing levels of CO_2 in the atmosphere but from the release of trillions of tonnes of methane trapped in clathrates, resulting from thawing of Arctic and Antarctic ice.[33] The impact is further exacerbated by the rate of temperature rise at the poles, which is nearly double that of the rest of the earth, and by the potent greenhouse effect of methane, which is even greater than that of CO_2. Methane **supersaturation** is already being observed in surface water in both the Arctic and the Antarctic, giving even more reason to leave the earth's last source of carbon-based fuel untouched.

Supersaturation: a state in which the amount of material in a solution exceeds that which can be readily dissolved.

The Risks of Nuclear Power: Fact or Fiction?

Nuclear fission: the splitting of an atomic nucleus into smaller nuclei.

Nuclear power involves **nuclear fission** of uranium and plutonium. The neutrons released by the initial reaction can trigger a chain reaction that generates large quantities of heat. The heat is then transformed into mechanical energy via a steam turbine and subsequently into electrical energy via an electric generator. Although it is not a renewable source of energy (the uranium and plutonium are consumed in the reaction and cannot be regenerated), nuclear power enables large-scale electricity production with virtually zero emissions. In fact, nuclear-generated electricity has been estimated to have a carbon footprint even lower than that of renewables. While nuclear power is currently being phased out in five countries, thirty countries are continuing its use, and thirteen are seeking to increase their nuclear infrastructure. With so many points in favor, why have we not expanded our entire energy infrastructure to be based on nuclear power?

For one, disposal and management of hazardous nuclear waste, particularly uranium, are serious concerns. Waste can be kept for a decade in on-site storage pools at reactor facilities; however, longer-term storage remains a challenge, with the safest option at present being geological isolation. New chemistry research is paving the way toward improving nuclear-waste-treatment processes.[34] Professor Polly Arnold of the University of California at Berkeley, for example, is conducting fundamental research on uranium's reactivity to help devise safer disposal and decontamination techniques.[35] Continued research and development into nuclear-waste treatment and containment will hopefully eliminate risks associated with long-term nuclear-waste storage.

On a practical note, the expansion of nuclear capacity may not be an effective solution to meet our short-term emission targets simply because of the time required to set up a new reactor site, which can range anywhere from five to twelve years.[36] Smaller modular reactors, with capacities ranging from 3 to 300 megawatts,* could help speed up the deployment process and bring low-carbon energy to the grid faster. Small modular reactors also offer benefits in terms of both upfront capital costs and safety. The US company NuScale Power, for example, designs nuclear reactors that occupy just 1 percent of the space of conventional reactors. Their simplified design eliminates many of the components that present inherent safety risks in conventional reactors, and incorporates additional safeguards.[37]

Public acceptance, however, may be the bigger driver of adoption of nuclear energy in different countries. Although exposure to nuclear radiation can have serious health consequences to both humans and the environment, the perceived risk of nuclear power is largely exaggerated. Without diminishing the weight of the estimated 4,000 deaths attributable to the Chernobyl nuclear disaster of 1986, it has been shown that the direct impact of the radiation on public health may have been less widespread and severe than previously thought.[38] With the exception of people located at the nuclear reactor site on the day of the accident, the population living in the contaminated areas received relatively low doses of radiation, comparable to background radiation levels.[39] The more recent Fukushima nuclear accident of 2011 in Japan rekindled public concern over nuclear power. Apprehension over consumption of contaminated seafood persists despite studies evidencing minimal health risks.[40] In fact, the use of nuclear power has been shown to prevent deaths and improve public health. A 2013 study calculated that nuclear power around the world has prevented more than 1.8 million air pollution–related deaths and, if used in lieu of fossil energy, could prevent up to 7 million deaths by mid-century.[41] While concerns over radioactive contamination are legitimate, we cannot ignore the fact that each type of energy

---

* Traditional nuclear reactors typically offer anywhere between several hundred to a few thousand megawatts of power, depending on the scale of the site.

generation presents its own set of unique health and environmental risks. Still, public sentiment about nuclear energy must be respectfully addressed if it is to be a viable energy solution in the long term. If anything, health and environmental concerns associated with GHG emissions need to be taken as seriously as those associated with nuclear radiation.

The global energy mix is gradually changing, with non-fossil sources, such as nuclear and hydroelectric power, deemed to represent half of the growth in energy supplied over the next two decades. However, we are currently experiencing a boom in natural gas. With a 1.6 percent annual growth rate, natural gas is now the fastest-growing fuel and is expected to overtake coal as the second-largest source of fuel by 2035. The growth of oil is currently 0.7 percent per year but is projected to decline slowly. The growth of coal is expected to drop even more sharply, by 0.2 percent per year, and will cease to grow in around 2025. The most rapidly growing source of energy, at 7.1 percent per annum, is renewables, which include wind, solar, tidal, and geothermal power. Their share of the primary energy mix is expected to be up to 10 percent by 2035, from 3 percent in 2015. Optimistically, through government policies and financial incentives promoting the use of non-fossil energy sources, this increase will continue.

In renewable energy, Europe currently leads the way, with non-fossil sources expecting to reach around 40 percent of the energy mix by 2035. The largest growth over the next twenty years in renewables, however, is envisioned by China, adding more than Europe and America combined. As solar and wind technologies become more attractive economically, their market share is anticipated to increase worldwide. Indeed, the dropping costs of renewables – the costs of solar and wind power have fallen by 85 percent and 50 percent, respectively, since 2010 – is the main factor driving their adoption. This price drop is owed to a combination of technological improvements and market-stimulating policies. A recent study analyzing the causes of cost reduction in photovoltaic modules, for example, showed that government policies to help grow markets accounted for nearly 60 percent of the overall cost decline.[42] The drop in costs has meant that renewable projects are already beginning to operate without government assistance. At the time of writing, solar farms were being built without subsidies or tax breaks in both Spain and Italy. China plans to cease financial support soon to some of its existing renewable infrastructure. In fact, we have reached the point where the cost of developing renewable infrastructure can no longer be used as an argument in favor of fossil energy. A 2018 study, for example, concluded that renewable electricity was already outcompeting oil on price and beginning to challenge natural gas.[43] From an economic standpoint, it no longer makes sense to argue in favor of expanding fossil infrastructure when renewables power is already outcompeting all other sources of electricity.

There is a component of systems-level change to electrifying our energy sector. Specifically, increased investment in electricity-drawing technologies tends to advance production of and research in energy-storage devices, which themselves assist in the development of a renewable energy infrastructure. Similarly, increasing the accessibility and affordability of renewably sourced electricity will, in turn, encourage adoption of low-carbon, electric-powered products. In this regard, many recent technological breakthroughs have also rejuvenated hope in a future energy economy built around electricity generated from renewable sources. The replacement of internal combustion engines by EVs is revolutionizing the transportation industry, which currently accounts for roughly 60 percent of all liquid-fuel consumption and 20 percent of all energy consumption worldwide.[44] In 2017 there were three million electric and plug-in hybrid vehicles on the road, a 50 percent increase compared to 2016.[45] That same year, there were a record 250 million electric bicycles in China.[46] The global market share of EVs is anticipated to keep growing given the falling cost of battery packs, the expansion of charging infrastructure, and continued policy support.

On this note, it is worth addressing some claims around the true emissions-reduction potential of EVs. The GHG emissions associated with any vehicle technology is not limited to burning fuel; that is, there are also emissions associated with the multitude of materials and processes – including battery production – involved in manufacturing a new car. We should always ask how the carbon footprint of EVs compares to that of combustion engine vehicles, taking into account all the emissions associated with manufacturing. For the most part, the majority of a car's GHG emissions emerge from the exhaust pipe. There is some evidence to suggest that simply employing more efficient combustion engines will have a greater impact on carbon-emissions reduction compared to introducing electric cars.[44] But, while increasing energy efficiency through carpooling and optimized engine technology is always beneficial, it should not be a reason to avoid the electric option. The true carbon footprint of any one vehicle is complex and depends on the local energy infrastructure and the emissions associated with its manufacturing, distribution, and ultimate disposal. Moreover, studies evidence that, unlike those of combustion engine vehicles, the emissions associated with EVs are only anticipated to drop as the carbon intensity associated with the energy and manufacturing industries declines and the share of electricity from renewables rises.[45,47] Moreover, EVs offer additional environmental benefits, such as the reduction of groundwater pollution associated with oils and lubricants. Simply put, it is never too early to purchase an EV.

Amid the excitement of the growing number of electric cars on the road, however, trucks, boats, ships, trains, and aeroplanes still constitute most of the transportation industry's fuel demand.* The question remains as to whether a large vehicle, bearing heavy cargo and traveling over long distances, can feasibly run on lithium-ion batteries, which remain the technology of choice for electric vehicles.

The answer, until recently, was said to be "unlikely." The gravimetric energy density (i.e., the amount of energy supplied per unit mass) of lithium batteries is much lower than that of hydrocarbon fuels. Large-volume vehicles and ships carrying heavy cargo would therefore require bigger batteries (or more batteries), and at some point the energy supplied by the battery could not satiate the weight of the cargo and of the battery itself. Nevertheless, decarbonizing large-transport vehicles is not an all-or-nothing scenario. The limitations of current lithium-ion battery technology should not hold us back from attempting to reduce emissions associated with this sector. Hybrid-electric aircraft, for example, in which an electric propulsion system is integrated alongside a jet fuel engine to improve the propulsive efficiency, can offer a means to reducing emissions during flight takeoff and landing.[49] Eventually, as battery technology improves, we can conceive of a future with fully electric aircraft.

Already the impossible is becoming possible, with the recent announcement that the first electric-powered autonomous cargo ship will be delivered in 2020, and will be rendered autonomous

---

* Aeroplanes are among the worst emitters in terms of CO_2. The combustion of jet fuel by aircraft account for 2 to 3 percent of all energy use–related CO_2 emissions. On top of the massive fuel requirements of aeroplanes, emissions released at higher altitudes are nearly twice as potent as emissions released at ground level. Cargo ships, by contrast, have the lowest CO_2 emission levels of any other cargo transportation method, generating fewer carbon emissions per tonne of freight per kilometer compared to those of barges, trains, and trucks. Still, the heavy fuel oil that cargo ships burn has a particularly high sulfur content, resulting in significant amounts of sulfur oxides and nitrogen oxides being emitted from ship smokestacks. Sulfur oxides pose a particularly high risk to human health and the environment, and the shipping industry accounts for 13 percent of these emissions annually.[48]

by 2022.[50,51] The maritime engineering firm Kongsberg is developing the battery system, electric drive, and all the components for seaborne transportation and autonomous navigation. The plan is to initially use the vessel as a substitute for land transport: the electric ship is anticipated to replace a total of 40,000 truck journeys every year.

In November of 2017, Tesla unveiled the Tesla Semi, the first heavy-duty, all-electric truck. It can reach an impressive 65 miles per hour going up a 5 percent grade (which is 50 percent faster than the average truck) and boasts a 500-mile range on a single charge.[52] Barely ten days later, aircraft maker Airbus, engine manufacturer Rolls-Royce, and electrical technology producer Siemens announced that they would join forces to design and build a hybrid-electric aeroplane. A prototype that would by ready to fly is anticipated for 2020 and could come to market as early as 2030.[53]

Although these technologies certainly give new hope for a sustainable future, they alone do not constitute the full solution to the electrification challenge of large transport vehicles. Decarbonizing and expanding the electricity grid will be necessary to power a future electrified fleet of aircraft, trucks, and ships. For example, a 2019 study found that a fleet of electric aircraft carrying out all flights up to a distance of roughly 1,000 km would form an equivalent of 0.6–1.7 percent of global electricity consumption.[49] In addition to expanding the grid to meet the new scale of demand, the extent of their adoption would ultimately depend on access to cheap renewable electricity. It has been estimated that renewable power would need to drop as low as four cents per kilowatt hour (kWh) to reach economic competitiveness with jet fuel, in the case of aircraft. Furthermore, in the absence of aggressive policy, there are usually considerable jumps to be made for a new technology or product to become competitive on the market, let alone to preempt a well-established industry. Owing to the long design and service times of aircraft, it is anticipated that the aviation sector will likely depend on the availability of liquid hydrocarbons for decades to come. Similarly, heavy-duty trucks, as well as maritime and road transportation, will likely continue to rely first and foremost on liquid hydrocarbon fuels in the near to mid future. Despite the growing uptake of EVs, the number of combustion engine vehicles continues to rise. For example, although the number of EVs on the road is expected to grow to 130–230 million by 2030, depending on the policy scenario,[45] the total number of motor vehicles on the planet is projected to reach two billion over this period.[54] So, while the growing market share of EVs is not insignificant, it will be some time before electrification takes over the better part of our transportation sector.

Clearly it is challenging for the transportation sector to replace the use of hydrocarbon fuels at the speed required to meet the emission targets set out by the Paris Agreement. However, with the ever-decreasing cost of renewable electricity and continued technological developments, it is possible to imagine a future in which all our power needs – including large transport vehicles – are fully met by electricity generated from renewable energy sources.

The rise in the use of electricity-based technologies and processes to replace hydrocarbon fuels clearly requires the continued development and expansion of a robust electricity infrastructure based on emission-free energy sources. At present the amount of GHG emissions associated with electrical generation varies greatly between countries, and even regionally, because it depends on the local energy infrastructure being used. The range varies from a mere 60 grams (g) of emissions per unit of electricity in Iceland, generated primarily from geothermal and hydroelectric power plants, to 1,060 g of CO_2 equivalent per unit in Australia, where coal still remains the dominant source of electricity generation.[55]

The technical challenge of relying exclusively on renewable energy every day, everywhere on the globe, comes down to the issues of transport, storage, distribution, and intermittency associated with renewable electricity. Technologies such as photovoltaic cells and wind turbines are highly dependent on climate and geography, and limitations to electricity storage make it non-trivial to meet demand at all hours of the day. Still, despite these obstacles, global energy projections show that growth in renewables is underway and will almost certainly continue to rise, especially with new technological innovation and sustained government support.

Although the shift toward non-fossil energy sources is certainly a positive progression, it remains a slow one. Too slow, as a matter of fact. If our emissions trajectory was the hill of a rollercoaster, we would currently be located somewhere between the point of steepest ascent and the maximum height of the hill. The sooner we reach the maximum height of the hill, the sooner we begin the descent. Unfortunately, emissions are unlikely to reach their peak before 2040,[56] which does not bode well for achieving our goal of zero emissions by 2050. Our current emissions trajectory results from the fact that, despite growth in renewables, overall fossil-fuel consumption continues to rise. At the heart of this problem lies the USD 5.3 trillion in subsidies that the fossil-fuel industry receives globally each year.[57,58] With around 50 percent of our legacy fossil fuels remaining in the ground, many in the industry see little impetus to halt extraction of oil, gas, and coal completely.

Drivers of Emissions

The continued rise in oil, natural gas, and coal consumption is heavily related to the growing demand for energy needed to satisfy the Western lifestyle. A sobering number is the 7,604 kg of oil equivalent that an average Canadian consumes every year.[*] This amounts to 8.4×10^4 kWh of energy, which can be given context by comparing

---

[*]   IEA statistics © OECD/IEA 2020.

Table 1. **Energy consumption** associated with various activities.

Activity	Energy (kilowatt hours)
Worldwide energy consumption in 2017	1.1×10^{14}
Worldwide renewable energy production in 2017	3.0×10^{12}
Bitcoin's annual electricity consumption [59]*	4.7×10^{10}
Canada's per capita electricity consumption in 2018	1.4×10^{4}
Round-trip flight New York–London on a Boeing 747[61]†	6.2×10^{3}
Canada's per capita electricity consumption in 1960	5.6×10^{3}
China's per capita electricity consumption in 2017	4.6×10^{3}
Barrel of oil equivalent	1.7×10^{3}
Driving 100 km in a Model S Tesla at 80 km/h[62]	1.5×10^{1}
Five hours of running a desktop computer with monitor and printer	1.0×10^{0}
Five hours of running a laptop	3.3×10^{-1}
Charging an iPhone 6 (a single charge)	1.7×10^{-2}

Unless otherwise specified, data was obtained from IEA statistics © OECD/IEA 2020.

* The idea that the digitization of our industries can lead to lesser environmental impact is largely a myth. Computing infrastructure consumes enormous amounts of energy to operate; for example, bitcoin alone uses more electricity than do some entire countries.[60]

† This statistic assumes the aeroplane to be 80 percent full. Most flights are not at full capacity.

it with the various processes shown in table 1. The worldwide demand for energy is expected to dramatically increase as the population grows to an anticipated nine billion by the end of the century.

In addition to our seemingly endless demand for energy, our fossil-fuel consumption is driven by our high consumer lifestyles. Beyond our driving a car and switching on the lights when we get home, every product we purchase and consume, from a new smartphone to a cup of coffee, has associated CO_2 emissions. The emissions associated with everyday phenomena can be surprisingly difficult to estimate because the true carbon footprint of any product includes the extraction or harvesting of raw materials, the processing or manufacturing steps, and the distribution and transportation of products – not to mention the further processing and packaging materials, which have their own associated carbon footprints. Furthermore, many of our favorite amenities are made, either directly or indirectly, from oil- and natural gas–derived products. Everything from sports equipment and cosmetics to crayons and chewing gum to aspirin and toothpaste has its origins in petroleum products. Some of our most critical commodities, such as ammonia-based fertilizers, without which current levels of food production would not be possible, call for fossil fuels (in this case, natural gas–derived methane) in their manufacturing process. Indeed, they are so intrinsically tied to our lifestyles and into the global economy that there is hardly a sector of modern society that operates independently of oil, gas, and coal.

So, despite the increasing market share of emissions-free-energy suppliers and technologies, our ever-rising demand for energy makes it difficult for new zero-carbon infrastructure to keep up. What does it take, exactly, to get our emissions under control?

CO_2 emissions can be analyzed and projected using the *Kaya identity*. The concept was first introduced by the current president of Japan's Research Institute of Innovative Technology for the Earth, Yoichi Kaya, who developed the identity in the early 1990s while he was a professor at the University of Tokyo. According to Kaya, the total GHG emissions can be expressed as the product of human population, gross domestic product (GDP) per capita, energy intensity (energy consumed per unit of GDP), and carbon intensity (CO_2 emissions produced per unit of energy consumed):

$$CO_2 \; emissions = population \times \frac{GDP}{population} \times \frac{energy}{GDP} \times \frac{CO_2 \; emissions}{energy}$$

In other words, an increase in population, GDP per capita, energy consumption, or fossil-fuel usage will result in greater CO_2 emissions. Transitioning toward energy sources that emit less, for example by replacing coal with natural gas, can reduce carbon intensity without the need to reduce population or consumption. This highlights a key feature of the Kaya identity: the decoupling of economic growth from carbon emissions.[63] In other words, it is possible to reduce CO_2 emissions without compromising economic growth.

We should, however, point out that the relative impact of each variable in the Kaya identity is subject to wild variation around the globe. An increase in population, for example, will have less effect on GHG emissions in countries where GDP per capita is low, compared to wealthier countries where it is high. As wealth and resources are not distributed equally around the globe, the Kaya identity does not offer much insight when each variable is taken as a global average. Economic inequality is so acute that the wealthiest 10 percent of global population account for nearly 50 percent of the total lifestyle-consumption emissions. Conversely, the poorest 50 percent are responsible for only about 10 percent of total lifestyle-consumptions emissions.[64] If you are reading this book, it is likely that your lifestyle emits 10–20 tonnes per year, upwards of ten times the emissions of an individual who falls within the poorest half of the world's population.[65] The carbon-emissions gap is so wide that even within the world's "emerging economies," such as China, Brazil, India, and South Africa, the lifestyle emissions of their richest citizens still fall far from those in wealthier countries. There is even a stark contrast in individual carbon footprints *within* countries, with South Africa and Brazil being the most extreme cases in which the richest 10 percent have a per capita carbon footprint ten and nine times that of their country's poorest half of the population, respectively.

Albeit this point is controversial, population size still has direct consequences on global energy consumption and carbon footprint. On the global scale, slowing population growth could provide roughly 16–29 percent of the emission reductions necessary to achieve our long-term climate goals, such as limiting warming to no more than 2°C.[66] However, stating that we simply need to "stop having children"[67] is not a particularly helpful message and grossly oversimplifies the complexity of population dynamics, not to mention that it ignores the dramatic variability in per capita consumption. Sudden shifts in a country's population pyramid, for example, can have unintended social and economic complications associated with an aging demographic.[68] Moreover, population-control policy is understandably a complicated and sensitive issue with a long history of forced sterilization targeted at marginalized peoples.[69,70] Selective focus on population instead of per capita consumption as the principal driver of emissions places unfair responsibility on the global south; after all, countries with the highest fertility rates tend to have the lowest per capita carbon footprint, whereas consumption in developed countries with declining birth rates has been the principal driver of climate change to date.[71] As Detraz writes in her book *Gender and the Environment*, "shining light on population is often done without simultaneously (or alternatively) spotlighting the connections between patriarchal economic processes, consumption levels, lifestyle choices, and environmental change."[72] Clearly, population control as a strategy for emissions mitigation must consider the multiplex economic, social, and political processes that drive global trends in birthrates.

Even the most aggressive population strategies, such as imposing a worldwide one-child policy, although undeniably beneficial in the long term (i.e., in the coming centuries), would offer little to no effect in mitigating emissions in the coming decades and therefore would not provide a quick fix for meeting our near-term emissions targets.[73] Furthermore, there exist highly effective alternatives to implementing radical population policies. Investing in family-planning programs and female education has been demonstrated to have a positive impact on the health of both humans and the environment in the short and long term. Family-planning programs have already been demonstrated as an impactful and cost-effective means to control fertility rates in many countries,[74] and women's education and empowerment can enable further control of fertility, as well as alleviate poverty and provide a more skilled labor force. Most interestingly, it has recently been suggested that offering universal general education, particularly at the primary and secondary levels, could be the key to enhancing climate-change adaption because it promotes social capital and reduces vulnerability to natural disasters.[75,76] Indeed, it turns out that teachers may be just as, if not more, important than engineers and policymakers when it comes to tackling climate change.

While the Kaya identity offers a helpful way of modeling different energy scenarios, it carries limitations and should therefore be interpreted carefully. Factors such as GDP per capita and energy intensity (i.e., consumption per GDP), for example, are not entirely decoupled from one another in reality. And, as mentioned earlier, when taking global numbers, the formula does not properly capture local information, regional variations, or the relationship between population growth and CO_2 emissions. The interplay between demographics, energy consumption, and social and economic development are non-trivial, to say the least.

What the Kaya Identity does illustrate, however, is that emissions can be reduced without compromising population or economic growth, by lowering energy and carbon intensity. Economic indicators and carbon-emissions data confirm that carbon intensity and economic growth are not always linked. From 2008 to 2015, for example, the United States reduced its emissions by 1.4 percent annually and yet also experienced economic growth by 1.4 percent annually.[77]

The question remains as to the rate and the extent to which energy and carbon intensity can be reduced. Informed and ambitious policy, together with focused investment in carbon-reducing technologies, is necessary to effectively reduce carbon emissions within the time frame required by the Paris Agreement.

A Race against Time

At the time of the industrial revolution, during the period of its development from 1760 to 1840, Europeans made a gradual transition from human-, horse-, and oxen-powered machines to machines powered by fossil fuels. The creation of this new fossil-based energy infrastructure took about a hundred years. The world's present fossil-based energy infrastructure, requiring trillions of dollars, was built over a century. Transitioning away from our current system to one that is based on carbon-neutral and renewable energy will take time and sustained investment. History tells us that such an energy transition is likely to take several decades or longer, possibly stretching to the end of the twenty-first century.

Given the bleakness of the situation, it is not surprising that certain advocates are calling for other approaches, such as **geoengineering**, to prevent the impacts of climate change. Although geoengineering might have once been reserved for the realm of science fiction, being deemed too audacious for real-world implementation, the approach is now being increasingly endorsed by academics, technologists, and climate advocates who believe that it is our only remaining choice for lessening the worst effects of climate change. In his 2017 *Joule* commentary, MIT professor emeritus John Deutch writes: "It is difficult to be optimistic that mitigation on its own will protect the globe from the consequences of climate change ... the world must urgently turn to learning how to adapt to climate change and to explore the more radical pathway of geoengineering."[63]

Geoengineering: planetary-scale interventions aimed at managing the greenhouse effect by directly modifying the earth's climate system.

Geoengineering strategies currently comprise two streams. The first includes technologies aimed at directly removing CO_2 from the atmosphere, and the second is based in solar-radiation-management methods to modify the **albedo** of the planet. Ideas for direct GHG removal include adjusting land-use management to enhance carbon sinks, and enhancing natural weathering processes to remove CO_2 from the atmosphere more efficiently. Solar-radiation management, however, involves more direct approaches, such as stratospheric **aerosol** injection. In this method,

Albedo: a measure of solar reflection off a surface, usually on a scale of 0 to 1.

Aerosol: a mixture consisting of solid or liquid particles suspended in a gas.

solar-radiation-blocking particles are put into the stratosphere to induce a reflective effect in a manner analogous to the release of gases during a volcanic eruption. Compared to carbon-dioxide-removal methods, stratospheric aerosol injection, once deployed, could potentially reverse planetary warming within a few years. A recent study estimated that a future program to deploy stratospheric aerosol injection could cost as little as USD 2 billion per year.[78]

Still, opponents of geoengineering often point to its unforeseen consequences and its lack of addressing the root problem of burning fossil fuels. Stratospheric aerosol injection, for example, requires additional injections as a result of particles precipitating, and the impacts are unlikely to be uniform around the globe. The IPCC's 2018 report *Global Warming of 1.5°C* also acknowledges the potential ecological and health risks associated with altering weather patterns and atmospheric chemistry; however, it concludes that geoengineering strategies, particularly stratospheric aerosol injection, could be used as a supplementary measure to meet the 1.5°C target.[79] The decision to include aerosol injection, however, is largely due to the absence of alternative technologies that are able to reduce emissions quickly enough.

Solar geoengineering research pioneer Dr. David Keith of Harvard University, though careful not to endorse the strategy wholly, points out that the evidence for solar geoengineering to reduce climatic hazards is strong and that, although doing it certainly carries risk, there are also risks to not doing it.[80] So, while a multinational geoengineering program remains unlikely in the near future, the possibility has become the subject of serious consideration and deserves attention by climate researchers as anthropogenic emissions continue to rise.

The urgency in reducing GHG emissions has never been more evident. Allowing temperature rise to exceed a mere half degree beyond the 1.5°C target poses major risk to all life on the planet. The frequency of extreme weather events, such as droughts, floods, forest fires, and extreme heat, would increase significantly, forcing hundreds of millions into poverty. A mere half degree over the limit risks doubling the number people experiencing water insecurity. Insect species vital to crops may become completely eradicated by a mere additional half-degree rise in temperature, creating global food scarcity. Coral reef ecosystems may not survive, marine life would sharply decline, and sea-levels rise could add an additional ten centimeters to coastlines.

Yet somehow the sheer scale of the issue, together with political barriers, has made the execution of solutions (at best) slower than necessary. Any serious strategy to combat climate change must involve understanding the reasons the reduction of emissions remains a challenge and what it will specifically take to meet the goals of the Paris Agreement. Do we have an existential problem that is too big, too complicated, too costly, and too political to solve?

KEY TAKEAWAYS

- CO$_2$ emissions should peak to no more than 30 Gt per year by 2030 (a ~45 percent decline from 2010 levels) and then need to be reduced to zero by 2050 for a decent chance of global temperature rising to no more than 1.5°C above pre-industrial levels.

- Fulfillment of every country's original nationally determined contributions set out by the Paris Agreement will not allow us to meet the 1.5°C target.

- Limiting global temperature rise to 1.5°C requires the rapid reduction of emissions across all sectors of the economy – energy, land, buildings, transportation, and industry. Enhancing atmospheric carbon sequestration (i.e., creating negative emissions) will be necessary.

- Globally, the energy sector, which involves the production of heat and electricity, contributes more emissions than does any other sector, with coal-based power generation being the single biggest contributor to global temperature rise.

- Despite the rise in adoption of renewable energy sources, and increased electrification efforts, fossil-fuel consumption continues to rise.

- It is possible to reduce GHG emissions without compromising economic growth.

- Population control is not an effective strategy for meeting our near-term emission targets.

From Space to Earth and Back Again

The sheer scale, complexity, and urgency of climate-change action is nothing short of overwhelming for anyone to process. We do not typically brace ourselves for uneasy emotions when reading a non-fiction science book; however, the seemingly inevitable arrival of climate change is hard to stomach. While feelings of helplessness are certainly founded, we believe that much of the fear and confusion surrounding climate change stems from the uncertainty that it presents. Knowledge can be a powerful tool to face these uncertainties. Education can help us overcome these fears by equipping ourselves with the science-backed knowledge needed to advocate boldly for ourselves, our communities, and our planet. In this chapter we will take a major step back (actually leaving the solar system!) to gain some perspective on how this whole mess with CO_2 and climate change started in the first place.

Back to the Beginning

Where did carbon dioxide originate? This question does not usually arise in discussions around climate change. To answer it, we must go all the way back to the big bang, when the universe was but a miniscule singularity, which then expanded during 13.8 billion years to create the cosmos as we recognize it today.

The start of the universe was marked by a rapid expansion of matter in a state of extremely high density and temperature, followed by a period of cooling, which led to the formation of subatomic particles and eventually atoms and molecules. At 10^{-43} seconds into the life of the universe, gravity became the first of the fundamental forces to distinguish itself. Before this time all four fundamental forces – gravity, electromagnetism, the weak nuclear force, and the strong nuclear force – are thought to have been unified under a single fundamental force. At 10^{-36} seconds, the temperature of the universe dropped sufficiently to differentiate the electroweak and strong nuclear forces. The electroweak transition followed at 10^{-11} seconds, whereupon electromagnetic and weak nuclear forces became distinct from one another. At one microsecond, protons and neutrons formed in the event known as the quark-hadron transition, and at three minutes, nucleosynthesis occurred, at which point light elements such as deuterium, helium, and lithium formed, defining the period of nuclear fusion. At around 5,000 years was the onset of gravitational collapse, characterized by the domination of matter throughout the universe. At 400,000 years, atoms began to form, and at 700 million years, the earliest visible galaxies emerged. It was not until 9 billion years into the life of the universe – that is, 4.6 billion years ago – that the collapse of a dense cloud of matter, whose swirling motion formed a rotating disk of gas and material, created our sun and solar system. Not long after the creation of our solar system, roughly 4.5 billion years ago, Earth and the other planets were formed by the culmination of matter in the outer reaches of the solar system. Although the earliest fossil evidence for life on earth dates to 3.7 billion years ago, the first life forms are estimated to have appeared as early as 4 billion years ago.

Hydrogen, helium, and lithium, the lightest elements, were formed during the early stages of the big bang, but other elements, such as the constituents of carbon dioxide – carbon and oxygen – were formed within stars by a process known as nuclear fusion.

All atoms are composed of a positively charged nucleus containing protons and neutrons, which are surrounded by negatively charged electrons. The high pressures inside stars are enough to fuse the nuclei of lighter elements together to form heavier elements. These reactions release energy and push material outward, counteracting the star's gravitational force, which is simultaneously pulling matter toward the star's center. As the fusion reactions continue, the nuclei become increasingly large and, at some point, become too large and stable to fuse further. With no more nuclear reactions forcing material outward, the star collapses under its own gravitational force, resulting in an enormous release of matter and energy, known as a supernova. This is the process by which elements are released into space.

As mentioned, the planets of our solar system are the result of accretion of solar nebula – that is, the collision and accumulation of the dust of particles spinning around the Sun and leftover from its formation. The earliest atmospheres of all planets were also made of solar nebula, which consisted mainly of hydrogen. However, the planetary atmospheres evolved over time, becoming incredibly diverse, and range from the thin and tenuous to the dense and powerful. The planets' atmospheres may consist of everything from hydrogen and helium to oxygen, nitrogen, carbon dioxide, ammonia, and methane. Interestingly, the only planets containing CO$_2$ in their atmospheres are Earth, Mars, and Venus. Mars has a thin atmosphere composed mostly of carbon dioxide, with smaller quantities of argon and nitrogen, along with trace amounts of water and oxygen. Venus, however, has a dense, carbon-dioxide-rich atmosphere with small of amounts of nitrogen and traces of methane and ammonia. It is therefore no surprise that Venus is the hottest planet in our solar system, with a mean surface temperature of 462°C, because its atmosphere creates a particularly potent greenhouse effect.

Earth is believed to have been formed around five billion years ago. A dense atmosphere emerged in the first 500 million years from the vapors and gases that were expelled during the degassing of the earth's interior. This early atmosphere consisted mainly of carbon dioxide, carbon monoxide, water, nitrogen, and hydrogen and was entirely free of oxygen.

Around one billion years ago water vapor in the atmosphere condensed, creating the oceans, which gave rise to the conditions that spawned the earliest aquatic organisms. These early life forms began to use energy from the sun to combine water and carbon dioxide photosynthetically into organic compounds and oxygen. Part of the oxygen that was created photosynthetically combined with organic carbon to re-create carbon dioxide. The remaining oxygen that accumulated in the atmosphere initiated a massive ecological disaster, at least from the perspective of these early existing anaerobic organisms.

At this time the oxygen in the atmosphere increased while carbon dioxide decreased. Some of the oxygen, located high in the atmosphere, absorbed the ultraviolet rays from the sun and created single oxygen atoms, which, upon reaction with oxygen molecules, formed ozone (O_3). Fortunately the ozone so formed was very effective at absorbing ultraviolet rays. It functioned as a thin shield surrounding the planet, absorbing wavelengths from 200 to 300 nanometers (nm), and thus protected the earth from biologically lethal ultraviolet radiation. Ozone is estimated to have come to existence around 600 million years ago.

At this time in the evolution of the earth's atmosphere the oxygen level was about 10 percent of its present concentration. With less ozone in the stratosphere, life was restricted to the oceans, which absorb sufficient amounts of ultraviolet radiation to allow aquatic life forms to proliferate. However, over time the increase in the concentration of photosynthetically generated oxygen, and the formation of correspondingly high levels of ozone, brought forth the first land organisms, marking the start of the evolution of life on earth.

Excavating Earth's Carbon Cycle

Carbon is one of the most abundant elements on the earth. Life on Earth is carbon-based. While plants, algae, and certain bacteria metabolize carbon dioxide to provide themselves with energy, other organisms convert it into an exoskeleton through a **biomineralization** process. This is exemplified by carbonate biominerals, which abound in calcite and aragonite coccolithophores (as shown in figure 5), as well as in sponge spicules, echinoderms, corals, and molluscan shells.[81] The shapes and patterns of these exquisitely sculptured biominerals continue to fascinate and deepen our understanding of **morphogenesis**.

Biomineralization: the process by which a living organism forms a mineral.

Morphogenesis: the processes by which the form or shape of an organism originates and develops.

Carbon pervades all biological materials, is present in geological formations, in the oceans as carbonate, and exists in the atmosphere as CO$_2$ gas. Carbon exchanges between these carbon sinks in a process known as the carbon cycle, as shown in figure 6. Although this process is largely balanced in a natural ecosystem, human activity has led to the addition of billions of tonnes of previously "stored" carbon to the cycle. According to climate scientists, roughly half of the annual production of CO$_2$ generated from fossil fuels is absorbed by the oceans and terrestrial biosphere, while the rest stays in the atmosphere. This is critical because it is a question of not only how much CO$_2$ is released, but also *where* the CO$_2$ ends up, that matters for maintaining the cycle's equilibrium.

The earth's carbon cycle consists of four principal reservoirs: the atmosphere; the oceans; the terrestrial biosphere, which includes all forms of life on land; and fossil fuels, which are the remains of plants and organisms buried underground. The interactions of atmospheric gases with the oceans and the terrestrial biosphere determine the distribution of CO$_2$ throughout these reservoirs, and

Figure 5. *Emiliania huxleyi*, type A coccolithophore.

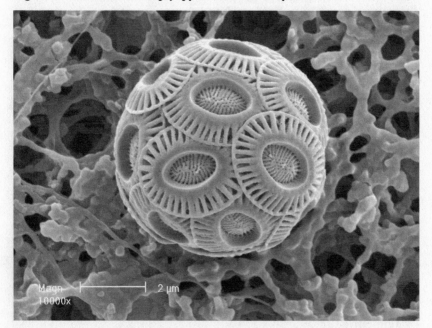

Reproduced from Young, J. R. *et al*. A guide to extant coccolithophore taxonomy. *Journal of Nannoplankton Research*, special issue 1, 1–132 (2003), with permission from Dr. Jeremy Young, University College London.

the ways in which these interactions change over time ultimately affect global temperature and climate. Understanding the dynamics and loads of various parts of the carbon cycle not only allows climate scientists to predict fluctuations in the composition of the atmosphere, but also helps to identify which mitigation, adaptation, and reduction strategies are most appropriate for addressing the associated climate change effects.

There are multiple pathways by which carbon moves between these reservoirs, with timescales ranging from fewer than seconds to millions of years. Atmospheric carbon is transported to land via rain, by reacting with water to form carbonic acid (H_2CO_3). This, in turn, reacts with rocks, in a process known as chemical

Figure 6. **Schematic showing the carbon pathways between fossil, terrestrial, oceanic, and atmospheric reservoirs under the natural carbon cycle**.

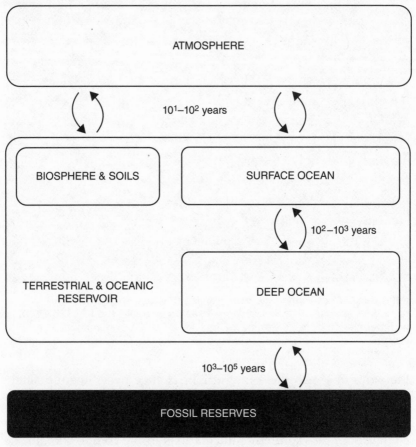

Adapted from Zeebe, R. E. History of seawater carbonate chemistry, atmospheric CO$_2$, and ocean acidification. *Annu. Rev. Earth Planet. Sci.* **40**, 141-165 (2012).

weathering, to produce various ions, notably Ca^{2+}. The ions are then carried away by rivers to the ocean, where they combine with bicarbonate ions (HCO$_3^-$) to form calcium carbonate (CaCO$_3$), carbon dioxide, and water. Carbon dioxide is also exchanged between the atmospheric and oceanic reservoirs at the surface of the ocean.

Whether the CO_2 gas molecules diffuse into or ventilate out of the seawater depends on the relative concentrations of CO_2 in the ocean and the atmosphere, as well as the solubility of the water. This interchange pathway is actually more effective at regulating atmospheric carbon dioxide in the short term (that is, on timescales on the order of less than 100,000 years), compared to the chemical weathering of rock pathway, whose feedback loop takes millions of years. Over the past two hundred years oceans have taken up roughly one-third of all anthropogenic carbon emissions. The sudden influx of atmospheric carbon dioxide has severely altered the equilibrium of carbonate species in the oceans, resulting in dramatic acidification and reduced carbon absorption by the oceanic reservoir.[82] The reason for this is that, when dissolved in seawater, carbon dioxide produces bicarbonate and hydrogen ions, which cause pH levels to drop. A 0.3–0.4 reduction in pH, equivalent to a 150 percent increase in concentration of hydrogen ions, has been projected for the world's oceans in the twenty-first century. Moreover, high concentrations of dissolved carbon dioxide shift the chemical equilibrium toward the production of bicarbonate ions (HCO_3^-) and away from carbonate ions (CO_3^{2-}), which consequently hinders the formation of calcium carbonate ($CaCO_3$), particularly in the deep sea.[83] As a result, shallow waters are becoming saturated with carbonate species, and the oceanic reservoir's CO_2 uptake efficiency is declining. The effect of rapid ocean acidification is expected to have such a dramatic impact on the oceanic carbon cycle and on marine biodiversity, particularly on calcifying organisms, that scientists have dubbed it "the other CO_2 problem."[84]

The formation of calcium carbonate in the oceans can take many forms: corals and phytoplankton, such as coccolithophores, as well as barrier reefs and carbonate banks. Eventually, when these organisms die, they sink to the ocean floor, where the carbon becomes stored within the layers of sediment. Similarly, carbon originating from organic matter, such as dead plants and animals, both on land

and in the oceans, becomes embedded in rock. The sediments are compressed over millions of years and eventually converted into petroleum reserves.

Most of this process took place in the pre-dinosaur age, 286–360 million years ago, during the Carboniferous period of the Paleozoic era. As partially decayed trees and plant material accumulated, they formed peat, which was gradually covered by minerals, sand, and clay, which in turn transformed into sedimentary rock. The growing pressure of the increasing weight of rock turned the peat into coal. Three types of coal were formed: anthracite, bituminous, and lignite. Anthracite is the hardest and contains the most carbon and the highest energy density. The softest and lowest in carbon, yet the highest in hydrogen and oxygen content, is lignite. Bituminous lies between anthracite and lignite. Tiny sea creatures called diatoms, which collected on the sea floor, constituted another source of fossil reserves. The increasing pressure of the overlying sediment and rock gradually converted the organic content of the diatoms into oil and natural gas.

In the pre-industrial world the only mechanism by which these geological carbon reserves could return to the atmosphere was through tectonic activity. During a volcanic eruption, carbon dioxide is released into the atmosphere, and lava and volcanic ash cover the surrounding land to produce a fresh layer of rock. In a business-as-usual scenario, the cycle's equilibrium would be maintained because the increase in atmospheric carbon dioxide triggers a rise in temperature, leading to more rainfall and subsequent rock weathering.

On a much faster timescale, carbon dioxide is exchanged between the atmospheric and terrestrial or oceanic reservoirs through photosynthetic and respiring organisms. Plants and phytoplankton take carbon dioxide from the atmosphere and, using sunlight, convert it to sugar. The process can be described in its simplest form by the following chemical equation:

$$6H_2O + 6CO_2 \rightarrow C_6H_{12}O_6 + 6O_2$$

This is a thermodynamically uphill reaction that needs energy to proceed; in nature this energy is provided by sunlight. The light sensitivity of photosynthetic systems results from pigments, specifically chlorophylls and carotenoids, whose chemistry has been optimized over evolutionary history to harvest visible spectral wavelengths from sunlight. Picture a rainbow, in which the range of visible light is broken into its components, starting with red, going to orange-yellow, followed by green roughly in the middle, and ending with blue and dark purple. The green color of grasses, leaves on trees, and other vegetation originates from photosynthetic pigments absorbing blue and red light, both of which are located at the ends of the range of visible light. With very little absorption in the middle of the visible light spectrum, under sunlight, grasses, leaves on trees, and other vegetation appear green.

The sugars produced in the photosynthetic process are the feedstock that enable plants and organisms to survive. Most photosynthetic systems create more sugars than they need to function, and the excess goes on to form carbohydrates, such as sugars, starch, and cellulose. This organic form of carbon can be returned to the atmosphere either through fire or by animals, such as humans, who consume the plant matter and exhale carbon dioxide. The exchange of carbon between the atmosphere and biosphere is so closely tied to photosynthesis that global levels of atmospheric carbon dioxide fluctuate with the seasons. Atmospheric CO_2 concentrations drop in the spring and summer months in accordance with plant growth in the northern hemisphere, and concentrations rise again in the fall and winter months, during which time many plants decay or cease to grow.[*]

---

[*] There is much more land and hence plant growth in the northern hemisphere compared to the southern hemisphere.

The fact that plants feed on carbon dioxide might leave you wondering whether rising concentrations of atmospheric CO_2 might actually be to their benefit. Will forests and crops not thrive in a more CO_2-rich environment? The effect of changing climatic conditions on photosynthetic organisms is, however, not quite so straightforward. While rising CO_2 levels favor photosynthesis, rising temperature levels risk having the opposite effect. The outcome of these competing forces, mixed with the many other factors affecting plant life, such as soil composition and weather patterns, remains challenging to ascertain. A 2019 study, for example, in which researchers modeled forests' response to climate change, concluded that averting significant drought and reductions in the earth's biomass was contingent on the ability of plant species to quickly acclimate.[85] In other words, whether forests would survive, let alone thrive, under higher atmospheric CO_2 concentrations is not clear. Even if plants respond positively to rising CO_2 levels, scientists are predicting that any benefits this may yield would be limited. For example, researchers at the University of Gothenburg in Sweden found that crops grown under elevated CO_2 conditions contained lower concentrations of nitrogen, potassium, and copper and therefore had significantly reduced nutritional quality.[86] Another study, this one focused on North America's boreal forests, found that, despite making trees use water more efficiently, higher CO_2 concentrations did not to lead more rapid tree growth and consequently did not enhance the amount of carbon they stored.[87] So, while higher CO_2 concentrations might favor photosynthetic processes, climate change, by its nature of interfering with well-evolved ecosystems, poses a serious threat to the quality of forests and crops.

Although we might tend to associate the terrestrial and oceanic reservoirs with plants and ocean water, animals themselves also contribute to the carbon-uptake capacity. Whales, for example, enhance carbon absorption in the oceans simply through the mechanical force of their diving, which carries nutrients up from deeper parts of ocean to the surface, and through their fecal matter, which helps to spread nutrients to new areas of the ocean, thereby attracting the growth of photosynthetic marine plants.[88] It is estimated that a single whale has the potential to capture thirty-three tonnes of CO_2 over its sixty-year lifetime, or the equivalent of one thousand trees.[89] The example of whales illustrates how conservation efforts have a major role to play in a long-term emissions-mitigation strategy. Biodiversity is critical to ensuring the uptake capacity of Earth's natural carbon sinks and maintaining the flows of earth's natural carbon cycle.

Recent findings have revealed the carbon-exchange process between the atmosphere and the earth to be more complex than previously thought. The National Aeronautics and Space Administration's (NASA's) Geostationary Carbon Cycle Observatory (OCO-2), launched in 2014, has been recording high spatial resolution data on the amount of CO_2 in the atmosphere and the uptake of CO_2 by the land biosphere. Such high-precision data has allowed researchers to identify, for the first time, the sources, sinks, and seasonal variability of CO_2 on both a regional and a global scale. Interestingly, they have found that the interannual variability in the carbon cycle, previously believed to be driven by the net sum of tropical vegetation on the planet, actually results from a combination of regionally specific effects, such as temperature, vegetation species, soil, and rainfall.[90–92] Continued monitoring of this kind will provide invaluable information to scientists trying to decipher the subtle complexities of the carbon cycle and ultimately the relationship between carbon and climate.

A Brief History of CO_2 Science

Our extensive knowledge of carbon dioxide and its role in the cycle certainly was not gained overnight. Like most topics in the physical sciences, it has taken centuries of incremental scientific inquiry and debate to understand carbon dioxide's chemical properties. The Flemish chemist, physiologist, and physician Jan Baptista van Helmont (1582–1644) is given credit for the discovery of carbon dioxide. At the time, in 1630, he called it wood gas (*gas* deriving from the Greek word for "chaos") because these were the vapors he observed being given off when he was burning wood. Scottish chemist Joseph Black later proved, in 1756, that carbon dioxide was present in the atmosphere, and resultantly called it fixed air. He also showed it to be a product of human and animal respiration and microbial fermentation, exemplified by the precipitation of limestone (calcium carbonate), by bubbling it into aqueous lime (calcium hydroxide), which could be reversed by heating the limestone. Around the same time, in 1754, Joseph Priestley discovered that carbon dioxide evolved from the action of oil of vitriol (sulfuric acid) on chalk (calcium carbonate) dissolved in water, which produced a pleasantly flavored, fresh sparkling soda. This discovery can be considered the foundation of the soft-drink industry. Humphry Davy and Michael Faraday, in 1823, discovered that, under pressure, carbon dioxide would liquefy. In 1835 Adrien Jean Pierre Thilorier demonstrated that the rapid evaporation of liquid carbon dioxide resulted in a solid flaky form of carbon dioxide, today known as dry ice. These discoveries offered the first glimpse into the fascinating chemistry of carbon dioxide and have enabled numerous modern technologies.

Tracing the Carbon Trail

With atmospheric CO_2 levels ever-increasing, understanding and characterizing the natural pathways of CO_2 into the atmosphere is a key task. Simply measuring the quantity of CO_2

in the atmosphere, for example, does not reveal whether the CO_2 came from the exhaust pipe of a truck or from the smoke of a forest fire. To understand the source of atmospheric CO_2, geochemists use a method known as carbon-isotope tracing. Carbon dioxide's three **isotopologues**, $^{12}CO_2$, $^{13}CO_2$, and $^{14}CO_2$, all undergo the same chemistry; however, their slight differences in mass result in their having different diffusion rates, as well as different **zero-point vibrational energies**. Heavier isotopologues diffuse more slowly, leading to variations of relative abundances of isotopes in different environments. The differences in zero-point vibrational energy also influence the energy barriers that must be overcome for certain reactions to proceed; that is, the same chemical reaction will occur at a different rate for different isotopologues. Heavier isotopes, for example, have higher energy barriers for bond-making and bond-breaking processes and proceed more slowly, despite being chemically similar to their lighter counterparts.

isotopologue: molecules having the same chemical composition but differing by an isotope. In the case of the CO_2 molecule, the carbon atom can contain either 6, 7, or 8 neutrons.

Zero-point vibrational energy: the characteristic lowest possible energy state of a quantum mechanical system. This is the energy that remains if all other energy sources (i.e., temperature, light) are removed.

Different carbon reservoirs have a distinct affinity toward the three different isotopologues and therefore can be used to trace the movement of carbon dioxide between the sinks. In photosynthetic reactions, for example, the lighter $^{12}CO_2$ isotopologue reacts more readily than the $^{13}CO_2$ does, and as a result the terrestrial biosphere has a preference for $^{12}CO_2$ over $^{13}CO_2$. Consequently there is a higher relative concentration of $^{12}CO_2$, compared to $^{13}CO_2$, in the terrestrial biosphere. The chemistry of the oceanic reservoir, however, does not prefer $^{12}CO_2$ over $^{13}CO_2$, resulting in a different relative uptake of atmospheric carbon dioxide isotopologues in the terrestrial and oceanic reservoirs.

Half-life: the time it takes for a radioactive substance to be reduced to half of its original amount.

Interestingly, while the $^{12}CO_2$ and $^{13}CO_2$ isotopologues are stable, $^{14}CO_2$ is radioactive, with a **half-life** of 5,730 years, meaning that its concentration changes with time. The signature exponential decay of $^{14}CO_2$ lies at the heart of radiocarbon dating, an invaluable tool that has allowed archaeologists, paleontologists, and geologists to determine the age of virtually any form of organic matter. Climate scientists can also use this property to identify the source of atmospheric carbon. As fossil fuels are formed over millions of years, they contain no $^{14}CO_2$ because the radioactive isotopologue has long since decayed to zero concentration. Monitoring changes in $^{14}CO_2$ in the atmosphere over time therefore offers strong evidence that the combustion of fossil fuels is responsible for the increasing levels of carbon dioxide in the atmosphere.

As a result of human use of fossil fuels since the industrial revolution, the total amount of CO_2 entering the atmosphere far exceeds that leaving it, as illustrated in figure 7. The carbon cycle's natural response pathways, particularly those of the oceanic reservoir, simply cannot cope with such pressures, and therefore cannot maintain their previous state of equilibrium.[93] Not only have excessive emissions caused the carbon cycle to be thrown out of balance, but climate change itself has already begun to alter the carbon uptake capacity of the other reservoirs.[94]

But how exactly is the earth's carbon cycle related to the planet's temperature? After all, the carbon cycle certainly is not the only natural system that we have disrupted on earth: human activity during millennia has permanently altered natural landscapes, shifted ecosystems, and destroyed entire species. What is it exactly about the carbon dioxide molecule that gives it so much power over the earth's climate?

You may have long been aware that carbon dioxide emissions from burning fossil fuels have enhanced the greenhouse effect. Chances are you learned how global warming works from Al Gore's 2006 film *An Inconvenient Truth*, which arguably marked a milestone in terms of bringing climate change into the popular consciousness.

Figure 7. **The excessive use of fossil fuels** since the industrial revolution has tilted the carbon cycle's natural equilibrium.

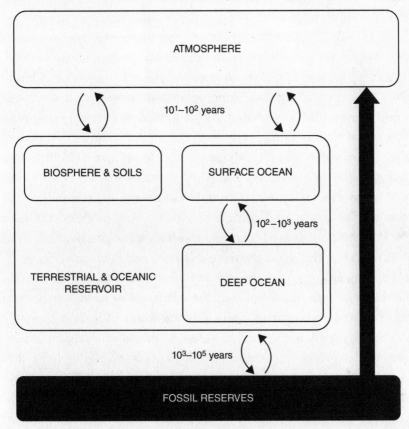

While Gore's explanation was correct, it does brush over some finer details pertaining to the chemistry of the CO_2 molecule in the atmosphere that may help answer some of these questions.

The Greenhouse Effect and Our Warming Planet

Our understanding of the greenhouse effect has a long history, and, like many scientific developments, its retelling often fails to assign credit to the correct individuals, largely thanks to the social constructs within which science and science history operate. In 1681, almost two hundred years before the world's first oil refinery was built, French physicist Edme Mariotte was the first to draw a comparison between the earth's atmosphere and a greenhouse. A century and a half later, in 1838, French mathematician and physicist Claude Pouillet presented the first correct analytical basis for the greenhouse effect, although his contribution is often miscredited to the more renowned mathematician Jean-Baptiste Joseph Fourier.[95] Together, Pouillet, Fourier, and English mathematician William Hopkins laid the groundwork that inspired Irish physicist John Tyndall's pivotal experimental demonstrations of the greenhouse effect. Tyndall wrote: "These speculations were originated by Fourier; but it was to M. Pouillet's celebrated Memoir, and the recent excellent paper of Mr. Hopkins, to which we were indebted for their chief development. It was supposed that the rays from the sun and fixed stars could reach the earth through the atmosphere more easily than the rays emanating from the earth could get back into space."[96] Although Tyndall is often credited with experimental verification of the greenhouse effect in 1859, it was earlier, in 1856, that American scientist and women's rights campaigner Eunice Newton Foote first discovered the ability of carbon dioxide and water vapor to absorb heat.[*] Foote demonstrated the basis of what we now know as the greenhouse effect through a series of experiments using an air pump, thermometers, and two glass cylinders, from which she observed that the heating action of sunlight changed under different gas environments. While Foote's experiments did not differentiate between direct solar radiation and the

---

[*] Foote's paper, *Circumstances Affecting the Heat of Sun's Rays*, was presented in August 1856 at a meeting of the American Association for the Advancement of Science, although Foote herself was not permitted to present the paper. The work was instead read aloud by Professor Joseph Henry of the Smithsonian Institution.[97]

infrared radiation that emanates from the earth's surface – the latter being the primary phenomenon underlying the greenhouse effect – they were the first to draw the link between the composition of the atmosphere and the temperature of the planet. Three years later, Tyndall, who likely was unaware of or had never read Foote's paper, reconfirmed Foote's results, using infrared radiation through a series of highly sensitive measurements that were made possible by his newly invented differential spectrometer.[98] Tyndall observed: "The solar heat possesses the power of crossing an atmosphere; but, when the heat is absorbed by the planet, it is so changed in quality that the rays emanating from the planet cannot get with the same freedom back into space. Thus the atmosphere admits of the entrance of the solar heat, but checks its exit; and the result is a tendency to accumulate heat at the surface of the planet."[96]

Infrared radiation: a type of radiation with energy higher than that of radio waves, but lower than that of visible light. Although it is invisible, we feel infrared radiation as heat.

Tyndall's description provides an excellent summary of the greenhouse effect: thermal radiation from the earth's surface, induced by the heat of the sun, is absorbed by GHGs and is radiated back toward the earth's surface, causing warming.

Nearly forty years later, in 1896, Nobel Prize–winning physical chemist Svante Arrhenius laid the groundwork for quantifying the greenhouse effect, and later, in 1938, Guy Stewart Callendar became the first to demonstrate that the earth's surface was warming; he even suggested that increased amounts of atmospheric carbon dioxide from burning fossil fuels were the cause.

In the operation of a real greenhouse, incident sunlight is transmitted through the glass and absorbed by the plants inside. As glass poorly transmits any reflected or emitted infrared radiation from inside the greenhouse, and because the heat generated is lost mainly by conduction through the glass, the temperature of the greenhouse rises. The earth's atmosphere behaves in an analogous way but with the additional heat-balancing effects of convective and evaporative cooling.

Still, what is it exactly that makes carbon dioxide – and not some other molecule – the principal culprit in this matter? The answer to this is surprisingly subtle and lies in the carbon dioxide molecule's intricate vibrational properties.

You probably encountered the electromagnetic spectrum at some point in science class. **Electromagnetic radiation** consists of everything from the microwaves that you might use to heat up your lunch to the high-energy X-rays used by doctors and dentists for imaging. It also includes visible light, that is, the part of the spectrum detectable by human eyes. What distinguishes all types of electromagnetic radiation is their energy. Radio waves and microwaves are low energy, whereas UV light, X-rays, and gamma rays lie on the high-energy end of the electromagnetic spectrum.

Electromagnetic radiation: energy-carrying waves that propagate through space, created by the synchronized oscillations of electric and magnetic fields.

Blackbody: an object that absorbs and re-emits all incident electromagnetic radiation such that it is in perfect thermal equilibrium with its environment.

To understand the origin and manifestation of the enhanced greenhouse effect, one needs to first understand what happens when electromagnetic radiation from the sun passes through the earth's atmosphere. The sun can be conceptualized as a **blackbody**; that is, it emits electromagnetic radiation solely as a function of its temperature. We recognize that the term *blackbody* is not intuitive; after all, the sun, far from being black in color, is the brightest object we know. However, designating an object as a blackbody simply means that the type and amount of radiation it emits is determined by its temperature. Hotter objects will emit more radiation than will cooler objects. For the physics curious, the relation between a blackbody's temperature and its spectral radiance is given by Planck's law.

When describing radiation that might be composed of many different energies, such as those that emanate from a blackbody, it is useful to show the object's spectrum, like the one shown in figure 8. The

Figure 8. **Solar irradiance spectrum** showing the spectral distribution of radiation emitted by the sun before the radiation enters the earth's atmosphere (white line), and the spectrum upon the radiation's reaching the earth's surface (light-gray area). The dark-gray area enclosed by the white line and the light-gray area corresponds to the incoming solar radiation absorbed by greenhouse gases.

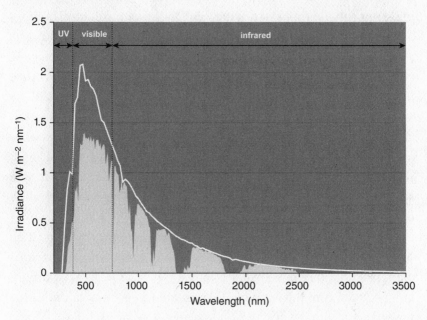

graph's horizontal axis corresponds to the frequency of the radiation; that is, how quickly the wave oscillates. The higher the frequency, the more energy the wave carries. The graph's vertical axis corresponds to the irradiance, which can be thought of as the intensity or the "amount" of wave of a particular energy. The spectra in figure 8 therefore shows the full breakdown of the sun's radiation: the wave energies of which it is composed, and their corresponding amounts.

As you may recall, the spectrum of a blackbody depends on its temperature. The sun's temperature of 5,525 kelvin (5,252°C) results in a spectral irradiance comprising about 51 percent infrared, 37 percent visible, and 12 percent UV, as shown by the white line in figure 8. This is roughly what the radiation spectrum looks like upon arrival at the earth's atmosphere.

The atmosphere is a layer of gases surrounding the earth; however, it can be further divided into four principal layers: the troposphere, the stratosphere, the mesosphere, and the thermosphere. The troposphere is the inner-most layer, lying closest to the earth's surface, and holds roughly 75 percent of the atmosphere's mass, including nearly all of its water vapor. The next layer, the stratosphere, is home to the ozone layer and also marks the highest altitude accessible to jet-powered aircraft. Next is the mesosphere, where meteors entering the atmosphere are typically burned up, giving the appearance of shooting stars. The outermost layer is the thermosphere, with which electrically charged particles collide to produce the stunning aurora borealis, and within which the International Space Station orbits the earth.

Our troposphere comprises mainly nitrogen and oxygen, with trace amounts of water vapor, carbon dioxide, methane, and nitrous oxide. From figure 8 we can see that roughly 12 percent of the solar spectrum is in the form of UV radiation (see the area under the white line to the left of the dotted vertical separating the UV and the visible regions). This is relevant because it is in this region of the electromagnetic spectrum that oxygen molecules absorb light, which causes them to split and form individual oxygen atoms. Other oxygen molecules then react with these oxygen atoms to form ozone (O_3), as described by this chemical equation:

$$O_2 + O \rightarrow O_3$$

The ozone molecules are a crucial component of the atmosphere because they literally shield us from cancer-causing UV rays. Back in 1974, chemists Frank Sherwood Rowland and Mario Molina showed how chlorofluorocarbon (CFC) propellants, under solar irradiation, can decompose in the stratosphere to form highly reactive species that cause the breakdown of ozone molecules back into oxygen.[99] In other words, they demonstrated how CFCs could accelerate the destruction of the ozone layer. So compelling were their findings that it prompted governments to fund further research into the problem. Ultimate validation of their theory came in 1985 when scientists Joe Farman, Brian Gardiner, and Jonathan Shanklin reported dramatic losses of ozone above Antarctica, exhibiting a veritable "hole" in the ozone layer.[100] The link between CFCs and the depletion of the ozone eventually led to the signing of the Montreal Protocol in 1987, which established an international mandate to phase out the use of ozone-destroying substances. Research has since shown that the treaty and its later amendments have been largely successful in slowly, but surely, remediating ozone depletion.[101] Despite this good news, the extent and speed of ozone layer recovery is now being called into question given the increased presence of GHGs in the atmosphere. Monitoring of the status of ozone in the stratosphere has become increasingly obscured by the effects of climate-change-induced changes to its composition and temperature.[102] So, although ozone remediation is underway, climate change is certainly not helping researchers track the degree of recovery.

The filtering of incoming solar radiation by the atmosphere can be appreciated in figure 8, in which we can observe a difference between the spectral distribution of solar radiation before it enters the atmosphere (shown by the white line) and that at the surface of the earth (shown by the light-gray area). The dark-gray area in between the white line and the light-gray area in figure 8 therefore corresponds to the combined absorption spectra of the various compounds found in the atmosphere.

The absorptive properties of the atmospheric gases result in roughly 70–75 percent of the sun's irradiance actually striking the earth. Depending on the type of terrestrial surface, some of the incident radiation is reflected back toward the atmosphere, though the majority is absorbed. The latter causes warming of the earth's surface, which, in turn, emits infrared radiation back toward the atmosphere. Note that while the sun sends a distribution of different kinds of radiation (UV, visible, infrared) to earth, the radiation that earth re-emits is mainly infrared. About 15–30 percent of this emitted radiation is sent into space, with the remainder being absorbed by the infamous GHGs in the earth's atmosphere; these in turn re-emit radiation toward both the earth's surface and space.

It helpful to think of GHGs as the insulation in the walls of a house. If the house is being built to withstand an extremely cold climate, you probably want it to have thick insulation to ensure that the house stays warm and cozy inside (conversely, if it is located in a region with a more moderate climate, thinner insulation will do just as well). The reasoning underlying these decisions is similar to the logic of the greenhouse effect. Sunlight shines through a window to warm the house and the heat slowly escapes through the walls. The insulation in the walls helps to control how much heat escapes back outside. Suppose that you are bracing for an extremely cold winter and decide to replace all the insulation with a thicker grade. The thicker insulation will result in less heat escaping the house. Having the same amount of heat entering, but less heat escaping will cause a shift in the system's energy balance, and the temperature inside the house will rise.

Just like our house analogy, any changes in the amount of incoming radiation to and outgoing radiation from the earth will cause a shift in its thermal equilibrium. If the rate at which heat enters the atmosphere was suddenly to become greater than the rate at which it escapes, the earth would get hotter, and its temperature would rise until a new thermal equilibrium were achieved. You may have heard of the term *radiative forcing* used by climate scientists. It is used to describe and quantify the factors influencing incoming and outgoing radiation to provide a measure of how far the planet's energy is out balance. If there were no radiative forcing, the balance between incoming and outgoing energy could be maintained and there would be no enhanced greenhouse effect. This, of course, is not the case at present, with an estimated 93 percent of the earth's energy imbalance being accumulated in the oceanic reservoir in the form of heat.[103] The continuous rise in ocean temperatures since the late 1950s is a clear indication that the global warming effect is well underway.*

Single-Layer Atmospheric Model

To appreciate better the effect of GHGs on the earth's temperature, let us take a look at what is known as the single-layer atmospheric model, illustrated in figure 9. Using only this simple diagram (and a little bit of mathematics), we can gain some insight into the impact of GHGs on the earth's mean temperature. Without going into the details, suffice it to know that the model comprises three main bodies that can radiate and emit heat: the sun, the earth's surface, and the earth's atmosphere. Assuming that the earth acts as a blackbody allows us to calculate the number of degrees by which the earth's surface temperature will shift with a given change in the atmosphere's **emissivity**. The emissivity of the atmosphere is a key parameter controlling the temperature of the earth's surface.

Emissivity: an object's capacity to re-emit energy as thermal radiation. It is defined as the ratio of radiation re-emitted by an object to that re-emitted by an equivalent blackbody object of the same temperature. An object can have an emissivity value ranging from 0 (zero emission) to 1 (blackbody radiation).

---

* The massive proportion of excess heat taken up by the oceans renders ocean temperature a more robust indicator than surface temperature for monitoring the anthropogenic influence on planetary warming because it is less prone to fluctuations and local variability.[103]

Figure 9. **Single-layer atmospheric model**.

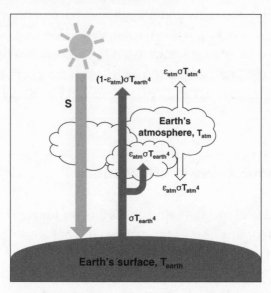

Reader Warning!

At this point the math-averse may skip ahead to the symbol ◊, which marks the end of the more technical section. Those interested in seeing the full model in action, read on.

Like most models used in physics, the single-layer atmospheric model rests on certain assumptions. The first assumption is that the atmosphere is transparent to incoming solar radiation, shown by the far-left arrow in figure 9, which is accorded intensity S. The second assumption is that the earth's surface acts as a blackbody, meaning that it absorbs all the incoming radiation from the sun and then emits it. Some of the earth's outgoing radiation is absorbed by the atmosphere, while the rest is transmitted directly into space, as shown by the dark-gray arrows in figure 9. Lastly, the model assumes that both the earth and the atmosphere are in thermal equilibrium, thus allowing us to employ an ideal version of Kirchhoff's law of thermal radiation, in which **absorptivity**, α, can be equated with emissivity, ε.

Absorptivity: an object's capacity to absorb radiation, specifically the fraction of incident radiation that is absorbed at a surface.

Recall that an ideal blackbody emits radiation simply as a function of its temperature. If the blackbody is at temperature T, the total energy it emits is given by σT^4, where σ is the Stefan-Boltzmann constant. Assuming the earth behaves as an ideal blackbody and that the earth's surface is at temperature T_{earth}, the radiation emitted by the earth's surface can be expressed as

$$\sigma T_{earth}^{4}$$

Unlike the sun and the surface of the earth, the model assumes that the atmosphere does not behave as a blackbody; that is, we will assume that it has an emissivity value of less than one. The amount of the earth's surface radiation absorbed by the atmosphere is therefore be written as

$$\varepsilon_{atm}\sigma T_{earth}^{4}$$

where ε_{atm} is the emissivity of the atmosphere. Similarly, the amount of the earth's radiation transmitted through the atmosphere and back into space is

$$(1 - \varepsilon_{atm})\sigma T_{earth}^{4}$$

Finally, the radiation emitted by the top surface of the atmosphere, toward space, and the bottom surface of the atmosphere, toward the earth's surface, can be written as

$$\varepsilon_{atm}\sigma T_{atm}^{4}$$

According to Kirchhoff's law of thermal radiation, the total energy entering the earth and its atmosphere, must be equal to that emitted back into space. Specifically, the incident energy from the sun must be equal to the energy emitted by the earth and the atmosphere. This can be expressed mathematically using the quantities defined previously:

$$S = (1 - \varepsilon_{atm})\sigma T_{earth}^{4} + \varepsilon_{atm}\sigma T_{atm}^{4} \tag{1}$$

Kirchhoff's law can also be applied to the atmosphere. The energy the atmosphere absorbs is equal to that it emits:

$$\varepsilon_{atm}\sigma T_{earth}^{4} = 2\varepsilon_{atm}\sigma T_{atm}^{4} \tag{2}$$

The factor of 2 in the above equation comes from treating the atmosphere as a slab that possesses both a top and a bottom surface. The atmosphere emits $\varepsilon_{atm}\sigma T_{atm}^{4}$ toward space, and $\varepsilon_{atm}\sigma T_{atm}^{4}$ toward the earth's surface, resulting in a total emission of $2\varepsilon_{atm}\sigma T_{atm}^{4}$. Solving equations (1) and (2) yields an expression for the earth's temperature in terms of the incident radiation from the sun and the emissivity of the atmosphere:

$$T_{earth}^{4} = 2S/\sigma(2 - \varepsilon_{atm})$$

The total radiation hitting the surface of the earth, which comprises the radiation from the sun and that emitted from the atmospheric slab, can be expressed as:

$$S + \varepsilon_{atm}\sigma T_{atm}^{4}$$

By substituting information from the previous equations into the second term and applying some algebraic manipulation, we get the following simplified expression for the amount of radiation hitting the earth's surface:

$$2S/(2 - \varepsilon_{atm})$$

From the expression we see that in the case of an atmospheric slab absorbing *all* of the earth's radiation ($\varepsilon_{atm} = 1$), the total radiation striking the earth is 2S; in the other extreme, where the atmosphere acts completely transparently ($\varepsilon_{atm} = 0$), it becomes S. In reality, the total radiative energy striking the earth is somewhere between these two limiting values. Taking S = 235 W/m^2 and $\varepsilon_{atm} = 0$ yields a planetary temperature of −19.45°C, while, for $\varepsilon_{atm} = 1$, the temperature rises to 28.6°C.

The atmosphere trapping some portion of the radiation emitted by the earth results in the earth's surface temperature being somewhere between −19.45°C and 28.6°C. These two extremes bracket the global mean surface temperature at 15°C.

$\Diamond$

From the simple single-layer atmospheric model we can quickly see the impact of the emissivity of the atmosphere on the temperature of the planet. The atmosphere's emissivity is largely determined by the absorption spectra of the atmosphere's constituent molecules. Understanding the infrared spectra of the individual GHG compounds is key because they largely dictate how much heat can be "trapped" by the atmosphere.

Greenhouse Gases

Although they constitute only 0.1 percent of the atmosphere's mass, GHGs play a critical role in determining the temperature of the planet because they effectively control how much heat is transmitted back into space. The main GHGs are water, carbon dioxide, methane, nitrous oxide, ozone, and CFCs. Their absorbing effect is illustrated in figure 10, which shows what the radiation escaping earth looks like before and after passing through the atmosphere. The white line

Figure 10. **Radiation spectrum of the earth** as seen from above the earth's atmosphere. The white line denotes the radiation emitted from the surface of the earth, and the light-gray area shows the radiation that goes back into space after passing through the atmosphere. The area located between the boundary of the white line and the light-gray area is effectively the portion of the radiation absorbed by greenhouse gases.

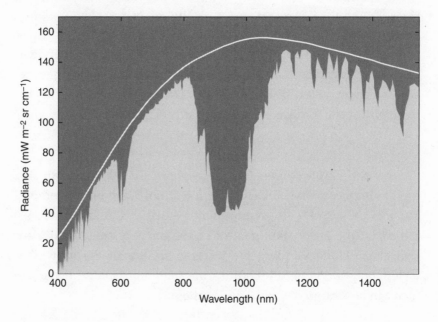

gives the distribution of infrared radiation initially emitted by the surface of the earth, and the light-gray area shows what the distribution looks like after the infrared radiation has passed through the atmosphere. The dark-gray area, enclosed by the white line and the light-gray area, therefore corresponds to the radiation absorbed by the GHGs. From this image we can appreciate the importance of understanding the absorptive properties of GHGs because these dictate the amount of radiation that escapes into space.

GHGs are characterized by their capacity to absorb infrared light. As mentioned earlier, the radiation coming to the earth from the sun is a combination of UV, visible, and infrared light. The radiation emitted by the earth, however, is mostly of the infrared portion of the electromagnetic spectrum so only the infrared-sensitive compounds in the atmosphere contribute to the greenhouse effect. Oxygen and nitrogen, which constitute most of the atmosphere's composition, do not contribute to the greenhouse effect, because they do not absorb in the infrared region. Still, some non-GHGs, such as carbon monoxide, nitrogen oxides, and sulfur dioxides, can affect climate, albeit indirectly. For example, a non-GHG can contribute to warming by reacting with other compounds to produce GHGs, by influencing the lifetime of a GHG in the atmosphere, or by affecting cloud formations that could in turn change the earth's albedo.

The ability of molecules to absorb and emit radiation originates from the motions of atoms within molecules. For this discussion, it is particularly helpful to imagine the atoms in a molecule as little balls connected by springs. The balls (atoms) can move, causing the springs (bonds) between them to stretch, bend, or twist. These motions are called molecular vibrations. Each vibrational motion has a characteristic frequency. When radiation impinges on a molecule its energy can be transferred to the atoms causing them to oscillate. This is analogous to the way in which the tap of the finger can set a bobblehead to wobble. A vibration is only activated if the radiation's energy matches that of the vibration's diagnostic frequency. As different molecules have different geometries (and therefore different characteristic modes of vibration), they absorb different parts of the solar spectrum.

Although all infrared-absorbing gases have the potential to contribute to the greenhouse effect, not all contribute to the same extent. After all, water vapor is a GHG, yet we do not hear environmental campaigns to cut our steam emissions. A common measure

Global warming potential (GWP): a measure of how much a greenhouse gas would contribute to global warming, in a given amount of time, if 1 kg were injected into the atmosphere compared to if 1 kg of carbon dioxide were injected into the atmosphere.

used to describe the impact of a GHG on the greenhouse effect is its **global warming potential (GWP)**. This measure attempts to quantify how much a GHG would contribute to global warming, in a given amount of time, compared to CO_2 (the GWP of CO_2 is one and serves as a reference from which the GWP values of other gases are calculated). A GHG's GWP value depends on the range of infrared radiation it absorbs, how strongly it absorbs, and the duration it resides in the atmosphere. GWP is expressed for a given period of time; for example, GWP-20 refers to a GHG's GWP relative to carbon dioxide over a twenty-year period. To give some perspective, methane has a GWP-20 of 84, meaning that would have 84 times more warming effect that carbon dioxide would over twenty years. Nitrous oxide, by comparison, has a GWP-20 of 264, and CFCs have a GWP-20 ranging from 6,000 to 11,000.[104] These numbers seem frighteningly high. Why is it that we are so concerned with CO_2 when it appears to be so much less potent? The reason lies in the fact that we have added excessive amounts of it into the atmosphere. Remember that the GWP considers the effect of injecting 1 kg of GHG compared to injecting 1 kg of CO_2. CFCs may be thousands of times more potent than CO_2; however, they only exist in trace amounts in the atmosphere.* The dramatic rise in atmospheric CO_2 makes it the single most important enhancer of the greenhouse effect. Methane is arguably the second biggest contributor, with its concentration having tripled since the industrial revolution; although methane and sulfur dioxide emissions are not the principal culprits, reducing them is

---

* Although emission of CFCs has stopped since they were phased out after the signing of the Montreal Protocol, they still do contribute to the greenhouse effect because of their very long atmospheric lives.

also considered critical to limiting global temperature rise. Emissions of halocarbons and nitrous oxide also contribute to warming, albeit to a lesser extent.

Another measure by which to compare the effect of GHGs, which has been increasingly adopted in recent years, is the **global temperature change potential (GTP)**. GTP is based on the change in global mean surface temperature at a chosen point in time of a gas, relative to that caused by CO_2. Just like with GWP, GTP values specified are for a certain period of time (e.g., 20 years, 100 years) and are with reference to CO_2, which is assigned a GTP of 1. Again, to give some perspective, methane has a GTP-20 of 67, nitrous oxide has a GTP-20 of 277, and CFCs have GTP-20 values ranging from 6,000 to 12,000.

Global temperature change potential (GTP): a measure of how much a greenhouse gas contributes to the change in global mean surface temperature, in a given amount of time, relative to that contributed by carbon dioxide.

Interestingly, water vapor is the most infrared-absorbing of all the GHGs. Its wealth of vibrational modes makes it absorb across almost the entire infrared region of the electromagnetic spectrum. Water, however, has been spared the bad reputation suffered by its fellow GHGs because the amount of water vapor in the troposphere, which is in large part governed by the earth's water cycle, has not been subject to change by human activity. Its exact residence time in the atmosphere is particularly challenging to measure due to the complexity of processes involved in the water cycle.

But water is an important part to this story. Although it absorbs a broad range of infrared radiation, it does leave a small window, located roughly in the middle of the infrared spectrum, through which radiation may escape directly into space. Unfortunately, part of the carbon dioxide molecule's absorption spectrum covers up part of this window. Carbon dioxide has four main vibrational modes, shown in figure 11, three of which are infrared sensitive.

Figure 11. **Vibrational motions of the CO_2 molecule**. The bending mode shown in schemas C and D are responsible for most of CO_2's greenhouse effect.

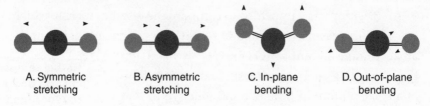

| A. Symmetric | B. Asymmetric | C. In-plane | D. Out-of-plane |
| stretching | stretching | bending | bending |

The symmetric stretching mode shown in schema A of figure 11 involves both carbon-oxygen bonds lengthening and contracting together, whereas the asymmetric motion illustrated in schema B involves one bond contracting while the other one lengthens. The vibrational mode shown in B is said to be infrared active because it can absorb infrared radiation. The symmetric stretching mode shown in A, however, is not. Schemas C and D show the CO_2 molecule's bending vibrations, both of which are infrared active.

Stare long and hard at the molecule's bending mode shown in schemas C and D. This is the vibrational motion that enables the absorption of energy at the infrared window that is left open by atmospheric water molecules. This specific vibrational motion of the CO_2 molecule is responsible for enhancing the greenhouse effect of our atmosphere and the global temperature rise.

Let us return, for a moment, to the analogy of GHGs as insulation on your house. Recall that adding CO_2 to the atmosphere has the same effect on the temperature of the earth as adding thicker insulation to your house. However, increased CO_2 concentrations also enhance the greenhouse effect in another way. For each vibrational state of the molecule, there exists a series of rotational states owing to the fact that the CO_2 molecule is not just bending but also rotating. As the concentration of CO_2 in the atmosphere rises, the probability of particles colliding increases. The additional energy gained upon a collision can activate these rotational modes, thereby increasing the molecule's capacity to absorb infrared radiation. In other words, higher concentrations of CO_2 in the atmosphere *widen* the range of radiation that can be absorbed. This is equivalent to both replacing insulation with thicker insulation and adding insulation to a wall of the house that was not previously insulated.

From its role in the global carbon cycle to its contribution to the greenhouse effect, CO_2 is clearly fundamental to our planet's chemical and thermal balance. It is rather incredible that such a small molecule could have such complex and far-reaching impacts on virtually all of earth's processes, especially when one considers that the environmental conditions under which all organisms evolved are the direct product of a specific atmospheric composition and an equilibrated carbon cycle. It is clear that any sudden, large-scale deviation in CO_2 levels will have inevitable consequences for all life on earth.

SF_6 Worries: The Most Potent Greenhouse Gas

It is not well known that the most potent GHG is, surprisingly, neither carbon dioxide nor methane but a colorless, odorless, and inert gas known as sulfur hexafluoride (SF_6).

Much like CO_2, SF_6 is among the most stable molecules on earth. The profound chemical robustness of SF_6 arises from the spatial congestion created by its six fluorine atoms' being tightly bonded about the central sulfur atom.

The high stability of SF_6 means that a tremendous amount of energy is released upon reaction, a property that has been utilized to great advantage in torpedo- and missile-propulsion engines. The unique properties of SF_6 also enable widespread applications that include dielectric insulation of electrical equipment, thermal window insulation, medical imaging, surgical procedures, semiconductor processing, metal casting, and audio equipment. The major problem with using SF_6 in all these applications, however, is the escape of the gas into the atmosphere during handling.

While its concentrations are only about 10 parts per trillion in earth's atmosphere, SF_6 possesses a GWP that is *23,900 times* that of CO_2 over a 100-year period. Already ten years ago, the contribution of SF_6 to climate change was estimated at 195 Mt CO_2 equivalents, and, owing to its high chemical stability, it can remain active in the atmosphere for up to 3,200 years, a worrying fact considering that its concentrations continue to rise.

It is worth mentioning that SF_6 is a unique GHG in that it does not exist in nature and is entirely a human creation. By contrast, CO_2 is a naturally occurring molecule, and although its current atmospheric levels are the result of human activity, its concentration has been higher at various times in earth's geological history. The consequence of rising SF_6 on our planet's chemistry, however, remains unknown.

Currently, electrical utilities and equipment are responsible for consuming 80 percent of the 10,000 tonnes of SF_6 produced every year, an amount that is growing with the increasing global production and demand for renewable energy. Unfortunately, growing use of SF_6 persists because there exist many technological and economic hurdles to switching to more environmentally benign alternatives. Although the withdrawal of SF_6 from service is argued to be costly, regulations are being imposed on both producers and users in an effort to minimize its use.

In Flight and Fight: Clear-Air Turbulence Caused by CO_2

The rising levels of atmospheric CO_2 are having repercussions well beyond the enhanced greenhouse effect, in aspects of modern life that we least expect. While it is common knowledge that aviation is a significant contributor to climate change through the CO_2 emissions from the combustion of jet fuel,[*] we do not realize the impact that climate change is having on aviation itself.

---

[*] Globally the aviation industry generates around 2 to 3 percent of anthropogenic CO_2 emissions.

New findings show that we now have reason to be concerned about the time we spend flying at altitudes of 30,000–40,000 feet. The latest research shows that high levels of anthropogenic carbon dioxide outside of the plane can intensify the jet stream in the tropospheric aircraft-flight paths, causing severe air turbulence.[105] To make matters worse, the resulting air movement is invisible to radar and therefore has real potential to compromise the safety of the crew and passengers. This intensification of clear-air turbulence by CO_2 stems from an increase in the temperature and density gradients between the equator and the poles that is induced by the combined effects of a warming troposphere and a cooling stratosphere.

Heightened levels of CO_2 are increasing the intensity and frequency of clear-air turbulence associated with more powerful jet-stream-related wind shears. Researchers believe that there will be a 40–170 percent increase in clear-air-turbulence incidences on transatlantic flight routes during the winter months if the concentration of CO_2 in the troposphere ever doubles pre-industrial revolution levels.[106] Already roughly 63,000 encounters with moderate to greater turbulence and 5,000 encounters with severe or greater turbulence occur globally every year. These incidences cause a large number of injuries to flight crew and passengers, and this can only worsen as the levels of CO_2 in the atmosphere increase.

As well as injuring passengers and crew, clear-air turbulence is responsible for many different kinds of structural damage to the aircraft. For US airline carriers alone, it has been estimated that the total economic cost of injuries, aircraft damage, flight delays, and time out of service because of repairs and pre- and post-accident investigations is in the range of USD 100 million to 200 million annually. All of these problems are expected to be exacerbated in the future because the aviation industry is anticipated to grow at a rate of 5 percent annually, meaning that it will double in about fourteen years.

In light of the anticipated intensification of clear-air turbulence, the current health and safety certifications of aircraft may not be sufficient to meet safe flying standards. To avoid encounters with moderate to severe turbulence pilots may try to evade them by adopting more convoluted flight paths. These inconvenient random motions lengthen flight times as well as boost fuel consumption and associated CO_2 emissions. All the effects lead to an increase in costs for the aviation industry in addition to the disruptions caused by climate change on the ground – such as temperatures being too hot for take-off[107] – increasing occurrences of lightning, and changing wind patterns at cruising altitude.

The good news is that this situation could improve in the next ten to twenty years as the dream of electrically propelled aircraft transitions from fantasy to reality thanks to improvements in battery technology. Prototype electric planes of early adopters are beginning to take to the sky.[108] There are already a number of companies vying in the competition for an electric vehicle, and we might be seeing airborne taxis in urban areas in the not too distant future.

KEY TAKEAWAYS

- Carbon can be found on land in both biological materials and geological formations, in the oceans as carbonate, and in the atmosphere as a gas. Carbon exchanges between these carbon sinks through a process known as the carbon cycle.

- The time scale of the earth's natural carbon cycle is too slow to regulate the sudden influx of carbon dioxide to our atmosphere that has been occurring since the industrial revolution.

- Excessive carbon emissions caused by human activity and climate change itself are decreasing the carbon uptake capacity of the oceanic and terrestrial reservoirs.

- Scientists use isotope tracing to track the sources of carbon. Specifically, monitoring changes in $^{14}CO_2$ in the atmosphere offers strong evidence that the combustion of fossil fuels is responsible for the increasing levels of carbon dioxide in the atmosphere.

- The saturation of carbonate in the oceans has led to ocean acidification, which is having dire consequences for marine biodiversity.

- Gases in the atmosphere insulate the earth by controlling the amount of radiation that leaves the planet. Radiative forcing involves any changes in the amount of incoming and outgoing radiation to earth. Shifting the planet's energy balance will trigger a change in planetary temperature.

- GHGs are characterized by their ability to absorb infrared radiation. The ability of molecules to absorb and emit radiation originates from the vibrational and rotational motions of atoms within molecules.

- The impact of a GHG can be quantified by its global warming potential (GWP) or by its global temperature change potential (GTP).

- Although all GHGs contribute to the greenhouse effect, the dramatic rise in atmospheric carbon dioxide makes it the single most important enhancer of the greenhouse effect.

- Increased concentrations of atmospheric carbon dioxide not only result in greater absorption of infrared radiation from earth but also curtail the atmosphere's natural infrared-transmission window.

Confronting Climate Change

From rapid extinctions of species to increasing ocean acidification to receding glaciers and to desertification, evidence of the changing climate is happening right before our eyes. The impact of human activity on the earth's biodiversity and geochemical signature since the time of the industrial revolution is now considered so substantial that it has warranted the debut of a new geological epoch, dubbed the Anthropocene. It is therefore particularly unsettling that in 2020, forty years after the first world climate conference in 1979, the technological challenges involved in decarbonization of the economy seem negligible when compared to the barriers posed by political and corporate interests. Any viable solution must be cognizant of how and why economic, political, and cultural forces might suppress CO_2-emission-mitigation efforts. These forces seem to be manifested largely by the principal attitudes of the following: the outright denier, the neo-skeptic, the cop-out, and the defeatist.

Attitudes in the Anthropocene

One attitude in the Anthropocene is associated with that of an *outright denier* who claims that either climate change is not real or it is real but cyclical and therefore not influenced by humans. Unfortunately, despite solid scientific evidence, this belief is still expressed among many in government, industry, and business circles. This head-in-the-sand view of global warming has inevitably affected the views of large segments of the public.

The *neo-skeptic* accepts the role of human activities in climate change and yet wields uninformed skepticism to deny the risks of climate change, questions the integrity of climate scientists, and diminishes the worth of mitigation efforts.[109] In facing neo-skeptical attitudes, the debate is shifted from the existence of climate change to the validity of response strategies. For example, neo-skeptics characteristically express concern for the economic and social consequences of adjusting our energy economy, often advertising the threat of job losses and irrecoverable economic demise. For them, the risk, uncertainty, and effectiveness associated with any proposed mitigation plan outweigh the consequences of no action at all.

At this point, let us reflect for a moment on two contrasting examples of urgent, yet seemingly impossible challenges that brought together groups of the brightest scientists and engineers working toward a common goal. The most prominent twentieth-century examples include the Manhattan Project and NASA's Apollo 11 lunar landing. These projects, although different in scale and scope, are unmistakably distinguished by the common associations invoked by humankind's most destructive and constructive potential. They express the conviction that anything is possible when we are willing and able to devote our full attention and work collaboratively to meet our most pressing challenges.

Climate change is, however, very different from the Manhattan Project and the Apollo 11 mission, which were concerted Western-led efforts motivated by warfare and geopolitical rivalry. Despite strong consensus among scientists around the world that anthropogenic emissions need to be reduced to zero, it remains challenging to meet targets when there is no silver bullet: emission-reduction strategies require a systems-level overhaul of the economy, from the local to the global. Moreover, physical climate models and anthropogenic warming predictions do face limitations. Even when climate-change policy is successful in reducing our emissions, there exists a strong possibility that the response of certain iconic climate quantities, such as mean global temperature, is slow and not in accordance with the initial predictions of models.[110] The uncertainty intrinsic to climate modeling can unfortunately be misinterpreted as grounds for labeling scientific studies as inaccurate, leading to confusion among policymakers and the public about the causes and reality of climate change.[111] This is particularly worrisome because in the majority of cases the uncertainty veers in the direction of underestimating – due to science's inherent conservatism – the rate, severity, and risk of global temperature rise. Therefore, while dissonance is expected of a healthy academic conversation, it creates many challenges to solving a problem that requires the full commitment of all involved. Anything short of a definitive and fully endorsed plan unfortunately creates the perfect conditions for neo-skeptical viewpoints to flourish. It is therefore important that climate scientists and economists heighten the exactness of decision science and the accuracy of risk analysis of the effects of a changing climate, and, most important, take care in communicating the relevance of their results to non-experts. Although more reliable information and better transparency might help to inform policymakers, the information-deficit model – that is, the assumption that the lack of understanding can be remedied by passing on knowledge – may not necessarily apply to climate-change skeptics. The relationship between knowledge, attitudes, and beliefs, particularly for politically polarized issues like climate change, is complex, and the communication of scientific fact is not a guaranteed approach to changing someone's view. Of course, this is not a reason for researchers and experts to cease engaging with the public; it just means that we cannot treat scientific-knowledge transfer as a guaranteed remedy to all false beliefs.

Defenders of a business-as-usual scenario, be they outright deniers or neo-skeptics, are often motivated by corporate interests and stand little chance of being swayed by refined risk analysis or further scientific evidence. This is because the scale of economic intervention that would be required to wean industries from fossil fuels stands in direct conflict to the current emphasis on quarterly profits that favors short-termism over long-term social, economic, and environmental well-being. The current state of the world economy has come at completely the wrong time for the transition from a fossil to a renewable energy economy because most business today is dictated by the logic of selecting the option that yields the highest **net present value**; that is, a benefit in the future is considered less valuable than the same benefit today.[112] However, such reasoning assumes a static profile of the resources and environmental conditions upon which society is dependent and does not necessarily consider the dynamic and unpredictable response of temperature rise on our air, water, land, and health. And yet vocal opposition toward a swift climate-change response persists under the pretext that aggressive intervention risks the collapse of the current economic system. Indeed, the slightest call to cease production or ease consumption is deemed to be a threat to the free market's requirement for continuous growth.[*,112] As put by Naomi Klein in her 2014 book, *This Changes Everything*, "changing the earth's climate in ways that will be chaotic and disastrous is easier to accept than the prospect of changing the fundamental, growth-based, profit-seeking logic of capitalism."[113]

Net present value (NPV): a method of calculating one's return on investment. It is computed by subtracting the initial investment from the present value of all future cash flows.

The third attitude posing a barrier to climate action is personified by the *cop-out*. Unlike the outright denier and the neo-skeptic, the cop-out is rarely motivated by financial interest; rather, the cop-out's

---

* Here, by *growth* we refer to economic growth, that is, simply the increase in gross domestic product (GDP).

attitude is adopted by those who acknowledge the existence of climate change, endorse climate action, and yet absolve themselves of any responsibility in the matter. This attitude impedes calls to action through the excuse that nothing can be done in the face of the systems (political or corporate) that are responsible for mitigating emissions.

The attitude also encompasses the belief that cop-outs do not need to engage in further climate action because they have already "done their part,"[114] as demonstrated in statements such as "I already recycle. What more can I possibly do?" While the individual or household may not have the power to change the emission intensity of global supply chains, we all have some degree of choice in the products we buy and the services we use. According to the IPCC's 2018 report, lifestyle choices that lower energy demand are "key elements" to limiting global temperature rise to 1.5°C. For example, investing in energy-saving appliances and devices can have a huge impact in reducing household energy consumption. After all, demand and supply go together in our current economy: a government's decision to invest in electric-charging stations requires the concomitant inclination of individuals to purchase EVs.

Of course, government also has a role to play in making greener products and services affordable and accessible to all,* and ultimately no meaningful impact can be made without swift institutional intervention.[115] In his book *How Bad Are Bananas?*, Mike Berners-Lee points to the limitations faced by consumers who wish to reduce their carbon footprint, and claims that "it's virtually impossible for an individual in the developed world to get down to a 3-tonne/year lifestyle anytime soon."[55]

---

* Interestingly, incenting energy-efficient options through tax programs and rebates is often not enough to shift consumer habits. The assumption that consumers (and organizations) are always inclined to make "rational" choices fails to capture the complex behavioral habits that need to be addressed to ensure that markets embrace energy-efficient technologies.

The fixation on the role of individual action in fighting climate change is not particularly constructive considering the scale of the energy and industrial sector's contribution to emissions. Indeed, *Carbon Majors Report 2017* revealed that a mere one hundred companies are responsible for 71 percent of all carbon emissions.[4]

Given the complexity and scale of the climate challenge, institutional-level transformation is clearly needed. But while it is impossible that widespread behavioral change alone will achieve our emission goals in the time required, denying one's own consumer power only risks slowing the market acceptance and adoption of new, low-emission technologies. Individuals have the capacity to induce institutional action by creating a culture of credibility and seriousness around climate change, which can in turn lead to changes in voting and even policy. We only have to look toward the examples of Greta Thunberg or Varshini Prakash to see how effective individual messaging can be. Furthermore, our individual commitment to becoming resource-conscious ancestors will matter in the long-term preservation of our environment.

The cop-out's attitude becomes more problematic at an organizational, corporate, or political level if responsibility and blame are being constantly attributed to other entities. Many countries with small populations that contribute less to global emissions may see action on climate change as futile, claiming that their own impact is negligible compared to that of more populous countries. "Why should we be willing to make sacrifices when the *real* problem lies outside our borders?"

The idea that responsibility for climate change should be relegated strictly according to the national emissions output is, at best, a gross oversimplification of our global reality. While it true that China, for example, contributes more emissions than any other country does, nearly one-third of these are from export production.[116] It is difficult to separate responsibility when we as consumers are inextricably tied to global supply chains.* More important, no progress can be made if everyone is waiting for others to take action. After all, why measure ourselves in terms of the problem when we could measure ourselves by how big a part we play in the solution? A comprehensive solution requires the coordinated efforts of individuals, markets, organizations, corporations, and government. We all have a role to play in the transition toward a sustainable economy and ultimately in meeting our collective emission targets.

The fourth and last type of attitude that stands in the way of effective action is that of the *defeatist*. As the name suggests, the defeatist is one who fully accepts the threat of anthropogenic climate change but sees no point in pursuing efforts because of the belief that there is nothing that can be done. It should come as no surprise that many of us have become hardened by the seemingly fruitless climate negotiations of the past thirty years and the impending environmental doom played out by our twenty-four-hour news media.[119] Most befitting in this regard are the observations of the late American author Neil Postman, who wrote that "most of our daily news is inert, consisting of information that gives us something to talk about but cannot lead to any meaningful action."[120]

It is particularly important that the outwardly alarming statements put out by the scientific literature, such as "[there is a]

---

* It is noteworthy that mass production of clean technology was largely a Chinese contribution.[117] It was China that instigated an unprecedented scale of skilled labor and manufacturing capacity dedicated to the production of wind turbines, solar panels, lithium-ion batteries, and water electrolyzer technology, an investment that was needed to lower the initial barrier to bringing renewable infrastructure to the global market. A 2019 study found that China's emissions are expected to peak at 13–14 Gt CO_2 per year between 2021 and 2025, approximately five to ten years ahead of the target set by the Paris Agreement.[118]

13% risk that committed warming already exceeds 1.5°C,"[121] and "less than 2°C warming by 2100 [is] unlikely,"[122] must not be misinterpreted by non-experts as validation of defeat. For this reason, science communication is key to the successful implementation of mitigation and adaption strategies. The public is entitled and should be encouraged to access accurate reporting of the latest scientific research on climate change, especially given its far-reaching implications. Effective science communication, however, requires selecting and presenting information in a manner that upholds factual integrity, while keeping to the objective of equipping non-experts with meaningful knowledge. In the case of climate change, informing the public of the gravity of the matter without painting a hopeless scenario can certainly be challenging, though not impossible. Dr. Matto Mildenberger, an assistant professor of political science at the University of California at Santa Barbara who studies public opinion, behavior, and policy around climate change, suggests that there is indeed evidence that overwhelming people with information can lead to a fatalistic mind-set. In a recent interview with the CBC he proposed an alternative approach: "The trick is to communicate the seriousness of the climate threat ... with a sense of empowering people to take action."[123]

On this note, we certainly do not wish to run the risk of triggering or stoking defeatism in the reader, particularly with this discussion on the challenges of overcoming climate-change inaction. Let us end this section with a positive thought: the earth's climate has not yet equilibrated from the particularly intense anthropogenic forcing of the past decades, which means that there is still opportunity to correct, or at the very least ameliorate, the earth's long-term response to the impulse of carbon emissions. In other words, we can choose to embrace a more dynamical view of the earth's systems and plan accordingly.

A Call to Action for Chemists

Chemists are largely responsible for inventing many of the key molecules and materials constituting the technologies that will enable the push toward a sustainable carbon-neutral economy. We all should be very proud of our scientific and technological discoveries that will help drive this transition. However, despite the many research contributions of the past decades, from advances in battery technology to the enhancement of solar cell devices, the fact remains that the general scientific community has been painstakingly slow when it comes to assessing prospective *gigatonne-scale* solutions to our emissions crisis.[124]

This is not meant as a criticism of science or to say that scientists have not been doing their job. After all, incremental advances in science are what drive technological innovation. However, researchers too must deal with the fact that we are no longer facing a business-as-usual scenario. Effective climate-change solutions require a scale, scope, and speed that are beyond any single technology or research discipline, which presents a major challenge to the scientific community that has been tasked with advising government and policymakers. Although many climate scientists work to inform and mobilize the public, news media, and government about the urgency of fast-tracking the transition from a non-renewable to a renewable energy economy, those of us in other research fields often remain quiet. Indeed, we tend to leave to others the equally important task of lobbying for climate change-mitigation. The introductions and conclusions of our papers are filled with motivations to help win the battle against climate change, and yet too often we do little more than publish the results of our research in scientific journals.

Little wonder that public surveys and polls reveal a growing skepticism and criticism of "alarmists," and a general distrust in even the most reliable data gathered and shared to date. It is our responsibility to correct this and to show how game-changing chemistry has the potential to drive the energy revolution. It is not enough to entrust lobby groups with doing our work for us.

We can show the impact of our work through worldwide outreach, high school and public education projects, and demonstration of the practicality of our innovations. These science communications must be highly visible and accessible to the public, otherwise pivotal contributions by the news corps will remain slow and sluggish, rather than fast-tracking our civilization to a sustainable future.

The time has come for those of us who passionately toil at laboratory benches to join the worldwide network of activists who are determined to make a difference. Our chemistry solutions cannot be implemented without us all working together to make them known, applicable, and effective. It is a commitment to our collective future that we all must make – to make it real.

Perception, Culture, and the Environment

The attitude of the defeatist is perhaps even more telling than that of the outright denier or the neo-skeptic because it reveals that, beyond technological and economic barriers, climate change remains a culturally challenging problem.[125] Calling into question what is arguably the most important resource sustaining our modern lifestyle, together with the sheer scale of the problem, renders climate change overwhelmingly complex to even the most well-intentioned individuals.[126] Perhaps the most difficult obstacle with which we struggle is the reality of perception: we look outside our windows and see a beautiful spring day, or we walk in open fields under blue summer skies, and can hardly believe there is anything wrong with our atmosphere that needs immediate care. As Tobias Haberkorn writes in his essay "The Coming Calamity," "climate change is too large for our senses ... The incongruity between the perception of one's own environment and that presented by the media is one of the most fundamental problems climate activists are facing."[127]

Changing everything from individual consumption habits to government policy in order to favor action on climate change requires that there be concern for climate in the first place. Individual concern for climate change and the environment is driven by a multiplex of factors, from location to income level to political ideology. Understanding the wealth and diversity in perspectives on the environment is crucial to designing effective, community-specific prevention and mitigation strategies in the short term, as well as effecting widespread behavioral changes in the long term. Much has been written on this topic, and in the following section we address a mere tip of the iceberg of how cultural, economic, and political processes contribute to the shaping of individual responses to climate change.

The human-nature relationship varies dramatically across the globe and is inherently tied to many factors including location, income, lifestyle, and culture. The 70 percent of humans who live in cities, for example, might not share the same awareness of the earth's changing land and climate as that of the world's rural farmers whose livelihood demands acute awareness of the weather patterns. Urban dwellers with enough income can readily access tropical produce in the middle of winter, order any item to their home by the click of a button, and leave trash out on the curb never to be seen again. We may at times observe the sudden rise in the price of a particular food, experience the mild discomfort of extreme temperatures in the winter and summer months, or occasionally lose power for a few hours after a severe windstorm; however, for the most part, we can, remarkably, live day to day without much concern for our local geography and climate, let alone appreciate any global-scale changes that might be underway.

Our disconnection from commodity chains, particularly for those of us living in the global north, shields us from seeing both the impact of our consumption on the planet and, arguably even more important, our own dependence on the planet's natural systems. The perception that we are somehow decoupled from our environment, though certainly fueled by our globalized economy, originates in the Western worldview that considers humans to be separate from the "natural" world and superior to all other species. This belief has been manifested throughout history, from colonialism all the way to modern-day capitalism, and has enabled justification and normalization of unabated extraction of the natural resources. The Western cultural viewpoint, however, is not the sole or standard perspective that societies have had on their environment. Anishinaabe* culture, for example, holds a uniquely intimate connection to the environment, best captured by scholar Lynn Gehl in her book *Claiming Anishinaabe: Decolonizing the Human Spirit*, in which she explains: "Inherent in an Indigenist approach ... is the need to care for, as well as the need to address the needs inclusive of all of Creation: sky, moon, water, trees, the winged, and the four-legged"; furthermore, "central to Indigenist knowledge philosophy is the appreciation of the limitations of the human-centered model of the world."[128] In fact, much of the language of today's mitigation approaches (such as "circular economies," "zero-waste technologies," and "intergenerational responsibilities") are mere echoes of concepts central to knowledge systems that have been forgotten or ignored or destroyed by Eurocentrism.

It might appear inevitable that the economy and culture that surround us are largely responsible for shaping our perspective on the environment and, in the present context, on climate change. The more we feel connected to or affected by our natural environment and climate, the more likely we are to take climate change seriously.

There are, however, less obvious social drivers, including political ideology, gender, race, and income level, that determine

---

* The Anishinaabe comprise a diverse collection of Indigenous nations across northeastern North America, including the Odawa, Saulteaux, Ojibwe, Mississaugas, Potawatomi, Oji-Cree, and Algonquin peoples.

individual perspective on climate change, albeit in ways not necessarily related to the logic of environmental connectedness. For example, research has found that men and women, in general, hold different degrees of environmental concern, with women expressing slightly greater concern than do men.[129] Women generally express greater concern for the health of their family and community, to which climate change poses a threat. The connection between political ideology and perceptions on climate change has also been explored by social scientists. McCright and Dunlap found that liberals (Democrats) and conservatives (Republicans) in the United States were exceptionally divided in their beliefs on climate change, with liberals being more likely to hold views that were more consistent with scientific evidence and to express greater concern for climate change compared to their conservative counterparts.[130] The authors of the study suggest that differing news sources and targeted political messaging are the main proponents fueling this growing cultural rift.

Concern about climate change may also be linked to risk perception. People who experience marginalization, whether it be through gender, class, race, or sexual orientation may respond differently to the threat of climate change than those who do not. Our perception of risk is often informed by our experience. Specifically, marginalized peoples may perceive a risk as being more "real" because when the threat presents itself, they do not expect adequate policy response.[72] In other words, people who have not experienced marginalization are more likely to trust that existing social systems will prioritize their safety and security in the event of a climate-change-related disaster; hence, they may be less concerned about the risk. On this note, in addressing the diversity of human perception of climate change, we cannot forget the hundreds of millions of people who have already been affected by one (or more) extreme weather events. The experiences of those who face disaster, from the family who lost its entire livelihood in a wildfire to the one that is without potable water after surviving a hurricane, will inevitably shape our species' ever-evolving view of the natural world. With the alarming frequency at which climatic disasters continue to strike, the portion of the world's population that falls into this category will, unfortunately, only continue to rise.

The varying levels of urgency for environmental protection and emission mitigation experienced by different peoples undoubtedly intensifies the social complexity of the climate-change problem. Recognizing the role of culture in informing individual knowledge and beliefs on climate change is critical to the successful implementation of mitigation and adaption strategies and must therefore not remain limited to the area of environmental philosophy.[125]

Climate Justice and the Road Ahead

When confronting climate change, we cannot ignore the fact that not everyone contributes equally to the net emissions caused by our species. Our consumption is determined by everything from income level to gender[*] to geography. Most of us in the global north undisputedly consume well beyond the minimum amount of energy required to achieve a decent standard of living[†] and a high level of well-being.[‡,133] Extreme weather events and resources scarcity do

---

[*] Worldwide there exist gender differences in energy consumption, with women consuming on average less than men. The difference originates from various factors, such as spending power, diet, and transportation habits.[131,132]

[†] The minimum threshold for an individual to achieve a good standard of living (that is, a life expectancy of above seventy years and full access to water, electricity, sanitation, and other infrastructure) is considered to be 30–40 GJ of energy.[133]

[‡] In their 1974 *Science* paper, Mazur and Rosa presented a compelling result showing that increasing energy consumption had no positive effect on quality of life beyond certain levels.[134] Their result holds true four decades later, with Mazur's latest work revealing that consumption beyond 40 MWh (144 GJ) per capita produces no increase in longevity in industrialized countries.[135]

not discriminate accordingly, and it is the most marginalized communities that will experience the greatest burden of the effects of climate change. The non-industrialized Inuit, for example, owing to the nature of air currents traveling from Asia and Europe toward the Arctic, are already suffering from some of the highest levels of pollution on the planet. Moreover, the non-localized climatic responses to fluctuations in the carbon cycle render it impossible for a single country, let alone a community, to manage and mitigate independently.[136] Our current organizational structures were not designed to accommodate phenomena that sporadically and unpredictably transcend geopolitical barriers. Who, for instance, should be held accountable when Texas faces dangerous levels of ozone due to British Columbian forest fires?[137] The spatial variability in temperature rise and weather, not to mention the very nature of wind and water currents, makes it exceedingly difficult to address climate change without the full commitment of all communities and nations.

Beyond national borders, climate change poses an intergenerational challenge, with the worst effects likely to be endured by those not yet born. At what point, then, do the risks imposed on future generations by our decisions become ethically unacceptable? These are non-trivial questions without obvious answers, which demand that we go beyond thinking of climate change as a purely environmental issue to be addressed with technological solutions. This is exactly where the concept of climate justice can be helpful. Climate justice is a framework that considers climate change as both an ethical issue and a political issue and demands that mitigation and adaption strategies be carried out in the spirit of promoting equity and upholding human rights.

Climate justice can be manifested in the form of legal action on climate issues, with individuals, households, cities, and organizations filing against both states and companies. This approach has proved to be successful in many instances worldwide, from the Netherlands[138] to South Africa[139] to the United States[140] to Norway.[141] Around the world, concerned citizens have taken different approaches: some lawsuits have targeted government for inadequate emission-reduction policy, and others have contested local industrial projects that pose serious climatic or environmental threats. At present, there are more than several hundred ongoing lawsuits worldwide related to climate change.

We should point out that although environmental activism is viewed by some as a radical pursuit in the West, it has been fundamental to survival in parts of the global south, particularly in tropical regions where livelihoods depend strongly on a thriving biodiversity.[142] Indeed, Ecuador became the first country, in 2008, to recognize the Rights of Nature in its constitution.[143] Since then, other countries, from New Zealand to India, have followed in granting legal justice to the natural entities, such as trees, rivers, and parks.[144] It was only at the United Nations (UN) climate talks in Bonn, in November 2017, that world leaders finally offered a long-overdue acknowledgment of the indisputable leadership of Indigenous groups in jump-starting response efforts to global climate change.[145]

Climate change is a global-scale problem that will not be resolvable without confronting current socioeconomic inequities and fulfilling our obligation to future generations. The situation in which we currently find ourselves presents a veritable opportunity to reshape unequal patterns in consumption because it already demands a high degree of international cooperation and rebuilding of infrastructure. It is most important to ensure that the market for luxury emissions by the wealthy minority does not outweigh the needs of the majority.[146] Lower-income countries, with less-established energy assets, could forgo altogether the West's fossil approach to development and leapfrog into renewable technologies, as they have done with mobile telephones.[147] Wealthier nations, in return, must do their part to assist lower-income countries financially in mitigating their emissions, and to take responsibility for managing the transition from fossil fuels while maintaining a stable and just economy.[148] In short, any viable solution, be it technological or political, must accommodate the welfare of all communities and nations.

Despite the many institutional, cultural, and ethical challenges in solving climate change, there are many reasons to remain optimistic. Individuals, communities, companies, institutions, and government around the world continue to prove that we have the capacity to overhaul our hundred-year-old fossil economy and mend our environmental damage, without sacrificing social well-being and economic prosperity. In Germany recent polls revealed that nearly 90 percent of citizens support the national *Energiewende* project, which seeks to transition rapidly into a low-carbon economy by adopting sources of renewable energy in lieu of existing coal plants.[149] In response to the exit of the United States from the Paris Agreement, businesses, cities, and entire states across that country have taken the initiative to maintain their original commitment of a 26 percent reduction in CO_2 emissions by 2025.[150] Where government and corporate initiative is falling short, groups of dedicated citizens worldwide are taking matters into their own hands.

We also know that consumer behavior has the capacity for rapid change, as demonstrated by the growing popularity of online shopping. This behavioral shift is already having far-reaching effects on advertising strategies, sales and marketing, consumer habits, and the environment. The research shows that social and behavioral changes in favor of climate action can be encouraged by policies that are as simple as the employment of more strategic advertising and consumer education. For example, offering credible and specific information at the point of decision can encourage climate-conscious behaviors.[151]

The physical reality of climate change is making it increasingly difficult for even the staunchest advocates of capitalist ideals to ignore the impact of extreme weather events on the financial sector. A 2016 study published in the journal *Nature* used financial modeling to estimate the effect of climate change on the world's financial sector. It reported that the total expected value at risk of global financial assets between 2015 and 2100 would be a shocking USD 2.5 trillion along a business-as-usual emissions path.[152]

In the same spirit, the United States' withdrawal from the Paris Agreement was decried by America's biggest companies, such as Microsoft, Apple, Facebook, and Walmart, for its anticipated economic repercussions, including dramatic loss of growth, jobs, and trade opportunities that participating nations would otherwise reap.[153]

Many governments are taking great strides to cut their carbon emissions, even at the expense of retiring their most enduring industries. In March of 2016, Scotland shut down Europe's largest coal-power plant, which had previously generated one-fifth of the country's carbon emissions, to become completely coal free.[154] In 2019 the United Kingdom became the first country to have its parliament declare a climate emergency. Although the motion does not legally compel the government to act, it sets a new standard for government response to climate change.[155] While language certainly does not equate to action, such a declaration from government can prompt industries to become serious about the prospect of radical decarbonization.

As we write this, the conversation around climate change continues to intensify. US congresswoman Alexandria Ocasio-Cortez and senator Ed Markey are forging a path toward national action on climate change with their introduction of a house resolution to create a Green New Deal that would see the decarbonization of the US economy within a few decades. What is novel about this proposal is that, unlike most environmental policies, which seek incremental progress through carbon-tax or -trade systems or targeted industry programs, it recognizes the need for a holistic approach in transforming the nation's economy into one that is both greener and more equitable.[156] In addition to achieving net-zero greenhouse gas emissions, its mission includes job creation and security, infrastructure and industry investment, access to healthy food and clean air and water, and the promotion of social justice and equity.[157] Initiatives such as the Green

New Deal in the United States and the Leap Manifesto in Canada*
provide excellent examples of the type of governance framework
that will be needed for countries to mitigate and adapt to the reali-
ties of climate change in a socially and environmentally just manner.

These are just a couple of anecdotes illustrating how the transition toward a low-carbon economy is no longer a predicted trend of
the future but a disruptive force that is well underway. The question
now is how this transition can be managed to meet emission targets
in an effective socially and economically responsible manner. It is
not beyond the realm of imagination, however, that these facts will
still fail to induce a response from those who are disillusioned by
short-termism and our fossil legacy.

Despite the voices of deniers, neo-skeptics, cop-outs, and defeat-
ists, movements pushing to meet our emission targets continue to
prevail. Our whole world needs to embrace that same optimism and
sense of hope that will enable our species to carry on for the next
hundred years, and the words we choose with which to speak will
inform the vision we choose to build for our planet.

Previous Extinctions and the Anthropocene

If climate change created mass extinction of life on earth, such a catastrophe would not be unprecedented; in fact, it would be the sixth
mass extinction of life on our planet. The previous five mass extinctions, originating from natural disasters on a massive scale such as
meteor impact or volcanic eruptions, managed to eliminate a significant fraction, or almost all, of the living species.

---

*   The Leap Manifesto, created by a consortium of Canadian authors, activists,
     and leaders in 2015, calls for a holistic restructuring of the Canadian economy,
     democratization of the country's energy sector, and the end of fossil-fuel use. See
     https://leapmanifesto.org/ for further details.

The historical record shows that the first mass extinction began around 400 million years ago, at the end of the Ordovician period. At this time the earth's dominant inhabitants were tiny sea creatures called graptolites. A sudden ice age drastically lowered sea levels and consequently wiped out over 85 percent of graptolite species. The reduced sea level contributed to significant drops in atmospheric CO_2 and a resultant lowering of global temperature. This may be attributed to the vast amounts of freshly exposed silicate rock reacting with CO_2, effectively removing it from the atmosphere, although alternative theories suggest that a sudden massive onset of plant coverage sucked significant quantities of CO_2 from the atmosphere via photosynthesis.

The second extinction began roughly 380 million years ago and saw the loss of nearly three-quarters of all species, the most abundant of which were the trilobites. Trilobites were a type of invertebrate marine animal ranging from a few millimeters to tens of centimeters in length. Their well-preserved exoskeletons left behind an impressive fossil record. Although the exact cause remains unclear, their demise was likely due to a sudden drop in oxygen in the oceans.

The third extinction, also known as the Great Dying, occurred some 250 million years ago and was – as its name suggests – the most devastating. It resulted in the loss of up to 96 percent of all marine species and 70 percent of terrestrial species. Notably, this extinction was caused by a temperature increase beyond 6°C, resulting from a lava flow in Siberia together with a concomitant release of methane from the permafrost. A 2018 study revealed that increased ocean temperatures and reduced oxygen levels were the principal causes of the extinctions.[158] It took about fifty million years for the earth's complex organisms to recover from this natural catastrophe.

Much later, some 210 million years ago, another extinction saw the loss of 80 percent of all species. Although the exact cause is not clear, it is thought to be an increase in atmospheric CO_2 from volcanic activity, resulting in global warming and ocean acidification.

Finally, the most recent extinction occurred around sixty-five million years ago, which completely ended the era of the dinosaurs and eradicated about half of all species on the planet. It coincides with the huge crater that forms Mexico's Yucatán Peninsula, which is also dated to about sixty-five million years ago.

Although the causes of the five mass extinctions were complicated, the annihilation of life was ultimately related to climate change. It was only about 2.6 million years ago that glaciation events, and associated climatic and environmental changes, caused the extinction of many species. Human activity began to have a noticeable effect on the rate of extinction of other species during the past twelve thousand years. The rapid release of anthropogenic fossil-fuel emissions into the atmosphere since the time of the industrial revolution has now blurred the distinction between planetary history and human history, and it has been proposed that we now find ourselves in a new geological epoch shaped by the human species: the Anthropocene.[159]

CO_2 on the Brain, and the Brain on CO_2

The adverse effects of high levels of CO_2 are well documented for space travel, scuba diving, submarines, aeroplanes, and firefighting situations. However, the impact of GHG emissions on human health from living in poorly ventilated environments is a disquieting surprise to many of us who are already worrying about the negative effects of GHG on our climate. Although we make and exhale CO_2, it is now apparent that we were not created to live in an atmosphere with increasingly high levels of CO_2. New research has shed light on the direct negative impact of CO_2 on cognitive functioning. We are not talking about the cause and effect of extraordinarily high levels of CO_2 but rather of concentrations that most of us would experience in closed, poorly ventilated spaces on a day-to-day basis, including homes, cars, and office and classroom spaces.

A public health study made in 2012 by Lawrence Berkeley National Laboratory and one published in 2015 by the Harvard T. H. Chan School of Public Health concurred that there were significant reductions in decision-making performance with increases in CO_2 levels from 600 to 1,000 to 2,500 ppm.[160,161] While these levels seem high in comparison to the 408 ppm of CO_2 currently recorded in our atmosphere, they are commonplace in poorly ventilated spaces of the kind mentioned earlier, where we spend most of our time. More recently, a 2019 study concluded that chronic exposure to CO_2 concentrations as low as 1,000 ppm could have adverse health effects, citing inflammation, reductions in higher-level cognitive abilities, and bone demineralization, to name a few.[162] These risks can potentially be mitigated through building-design measures that prioritize indoor air quality. Green residential buildings have, in some instances, been shown to have lower CO_2 levels compared to those of other residential buildings.[163] Fortunately there is push for the design of green buildings that optimize conditions for human health and productivity in addition to energy efficiency.[164]

Whether it be through extreme weather events, resource scarcity, or CO_2-concentrated air, climate change has a direct impact on human health. As such, it is crucial that our health-care systems be adapted for climate resiliency and that government and health-care professionals promote policy and behavior to reduce the health risks of climate change.[165]

KEY TAKEAWAYS

- Any viable climate-change solution must consider how and why economic, cultural, and political forces might suppress emission-mitigation efforts.

- The uncertainty that is intrinsic to climate modeling should not be interpreted as grounds for labeling scientific studies as inaccurate.

- Effective education on climate change requires communication of the seriousness of risks associated with temperature rise, while also instilling in people a sense of empowerment to act.

- Widespread behavioral change alone will not achieve our emission goals in the time required; however, denying one's own consumer power only risks slowing market acceptance and adoption of environmentally favorable technologies.

- Understanding the diversity of perspectives on the environment is crucial to designing effective, community-specific prevention and mitigation strategies in the short term and effecting widespread behavioral changes in the long term.

- Climate change is an ethical and political issue that cannot be resolved without confronting current socioeconomic inequities and fulfilling our obligation to future generations.

- The transition toward a low-carbon economy is no longer a predicted trend of the future but a disruptive force that is well underway.

4

Stubborn Emissions

At this point you hopefully have a decent idea of carbon dioxide's role in climate change, the latest trends in our energy economy, and the challenges we face in reducing emissions to meet the goals laid out by the Paris Agreement. We can see that the market share of renewable energy is steadily rising, and government and industry are increasingly facing pressure to decarbonize. Yet, at this rate, we are likely to fall severely short of meeting the emission targets required for a chance to keep global temperature rise below 1.5°C. The transition away from fossil fuels is simply too slow.

But even the most die-hard decarbonization advocates among us might agree that going cold turkey on fossil fuels would require unrealistically radical lifestyle changes. For example, let us consider the most optimistic case in which the entire planet's electricity demands could be sustained by renewable energies. What would the world look like if we were to immediately stop using all hydrocarbon fuels?

For one, global food security would be severely threatened due to the termination of fertilizer production and a shutdown of most freight transportation. Jobs would need to be replaced overnight for everyone working in the manufacturing, agriculture, and transportation sectors, let alone those working directly in the fossil-fuel industry. Commodities that required petroleum products, including many essential chemicals and pharmaceuticals, would be discontinued until new syntheses were discovered. Most of the population, in both high- and low-income countries, would be forced to abandon their gasoline-fueled cars and motorcycles. All commercial air travel would come to a halt. Beyond the social and political barriers described in chapter 3, the speed of the transition to an emission-free world is clearly limited by certain "stubborn" emissions of the industrial and energy sectors. To understand what makes these sources of GHG emissions so stubborn, we need to look in detail at their source.

The Devil Is in the Details

In chapter 1 we briefly described how industrial and chemical processes typically generate GHG emissions in two ways. However, if we are to be specific, they do so in *three* ways. The first is by the burning of fossil fuels to generate the high temperatures needed by many chemical processes. These emissions are associated with delivering energy to the process. This can be thought of as the energy required to heat your oven when you are baking bread. For example, the production of ammonia, the principal building block in all the world's fertilizer, requires temperatures as high as 450°C. When considering the scale at which many of these chemical products are made, one can quickly appreciate their planetary impact from heating alone. The chemical industry currently consumes roughly 10 percent of the world's energy, most of which is derived from burning fossil fuels. Emissions associated with supplying energy (i.e., heat) to industrial processes can be eliminated by using heat that is derived from renewable energy sources. Burning fossil fuel for heat might be convenient and economically favorable but certainly not necessary.

Feedstock: a chemical or material used in a manufacturing process.

The second way in which chemical manufacturing emits GHG emissions is through the production of the **feedstock** chemicals. This aspect of chemical manufacturing is not trivial: while energy accounts for roughly 40 percent of the fossil resources used in the chemical industry, the remaining 60 percent are used as chemical feedstock.[166] In our baking analogy, feedstock chemicals are equivalent to the ingredients in a recipe. Many of these ingredients may have associated emissions. For example, the flour originates from wheat or other starchy plants and undergoes multiple steps, including washing, drying, grinding, processing, and maturing before it is ready to be used. Most of these steps require some form of energy (e.g., heat for drying, and mechanical energy for grinding) and therefore are likely to have associated GHG emissions. In the world of chemical manufacturing there are certain feedstock chemicals that are widely used for a variety of processes, much like flour is a common ingredient found in most baking recipes. One of the most common "ingredients," or feedstock chemicals, is hydrogen gas. Unfortunately the most common method of producing hydrogen on an industrial scale, **steam methane reforming**, is extremely energy intensive. In steam methane reforming, methane (natural gas) and water are brought to extremely high temperatures and pressures to make **carbon monoxide** and hydrogen. This process itself requires a great deal of energy to reach such high temperatures and requires natural gas, a fossil fuel, as a chemical input. Fortunately this is not the only known way of producing hydrogen gas. Hydrogen can be produced from the **electrolysis** of water, which involves using an electric current to split water into its components, hydrogen and oxygen

Steam methane reforming: a chemical process in which syngas, a mixture of carbon monoxide and hydrogen, is produced from hydrocarbon fuels, typically methane (natural gas) and water.

Carbon monoxide: a colorless, odorless, and poisonous gas consisting of one carbon atom and one oxygen atom. It is widely used in chemical manufacturing.

Electrolysis: a process in which direct electric current is used to drive a non-spontaneous chemical reaction. Electrolysis of water involves using electricity to decompose water into hydrogen and oxygen gas.

gas. Not all chemical feedstocks, however, have an alternative way of being generated, many being fossil-derived hydrocarbons themselves. Therefore, employing electricity generated from renewable sources can potentially address some, though not all, of the emissions associated with producing chemical feedstocks.

The third way in which chemical and industrial processes can produce emissions is by the nature of the process itself; that is, carbon dioxide can be a by-product of an industrial reaction. This is equivalent to the release of carbon dioxide that happens when dough is baked. At this point you might be wondering, "What is carbon dioxide doing in my dough?" Without our going into full details of the science of bread making, many bread recipes require that the dough rise before being baked. At the molecular level the rising occurs because yeast in the dough consumes sugars and converts them into carbon dioxide, creating little pockets of gas that cause the dough to expand. The carbon dioxide is then released when the dough is baked. These carbon emissions have nothing to do with the energy required by the process; rather, they are intrinsic to the chemical reaction of baking bread. Similarly certain industrial processes result in carbon emissions that are by-products of the reaction. They result from the chemical nature of the reaction, rather than from the need to produce energy. This third type of GHG emission, common to the industrial sector, cannot simply be eliminated by employing renewably-sourced energy.

Using electricity derived from renewable sources can eliminate the carbon emissions associated with the heat and the hydrogen required for many chemical and industrial processes. Electricity can replace fossil fuels to generate heat, and electrolysis can provide a sustainable, emissions-free way of producing hydrogen. We should point out that, even though a solution exists, it does not mean that it can be implemented overnight. Practically speaking, we must consider priorities when it comes to implementing low-carbon energy. To start by replacing the most carbon-intensive pathways with low-carbon or zero-carbon solutions would be the most effective means in terms of emission mitigation. Replacing our fossil-based industrial sector will require a reinventing of much of our existing infrastructure: electrolyzers will be needed to generate hydrogen gas, industrial reactors will have to be redesigned and rescaled to accommodate electricity-derived power, and massive amounts of low-carbon electricity will need to be generated and made accessible. The main challenge lies in the latter point – the scale of low-carbon electricity that would be required. This is not to say that electrifying our chemical industries is impossible – it is technically feasible – but it is not yet achievable under current trends in renewable-electricity production. Overcoming this challenge will require large investment by stakeholders and aggressive policy from governments to kick-start its development.

There remain those "stubborn" emissions that are either associated with certain chemical feedstocks or intrinsic to the reactions themselves. These are harder to address from a technological standpoint and require a rethinking of certain industrial processes altogether. The industry's non-trivial reliance on fossil resources, and its resulting emissions, is the reason that chemical manufacturing is anticipated to be the single largest source of oil consumption by the year 2030. Although we have focused much of this discussion on the industrial sector, the transportation sector also has a stubborn emissions problem, as mentioned earlier. As we saw in chapter 1, electric vehicles are steadily occupying a greater share of the market; however, despite demonstration of electric heavy-transport-vehicle prototypes – and commercial adoption in some instances – they are unlikely to reach the commercial scale required to meet emission targets. It will be difficult to cut completely our usage of hydrocarbon fuels.

The key point we wish to illustrate here is that even with the most aggressive investment in renewable energy, which itself is not without its own set of challenges, it would be next to impossible to render our existing industrial processes and transportation usage completely emissions free. Generating electricity from renewable sources, despite being effective and necessary in the battle against climate change, alone cannot enable a rapid transition to a zero-emission economy. There must be other ways to help us achieve net-zero emissions.

To Remove or to Capture?

The concentration of CO_2 in the atmosphere is dictated by the amount going in, as well as that going out. Most of the conversation around climate action tends to focus on stopping further emissions from entering the atmosphere, and rightly so. After all, anthropogenic emissions are something our species should be able to control. However, is it possible that we have reached a point where we need to start seriously considering removing emissions from the atmosphere or capturing stubborn emissions right at their source? If we removed more carbon dioxide from the atmosphere than we put in, could this grant us additional time to transition our economy? And if we were to remove it, where would it go?

The next section will address some of these questions. First, let us clarify some terminology. Removing CO_2 versus capturing CO_2, although similar, are slightly different concepts. Carbon dioxide removal (CDR) involves removing carbon dioxide from the atmosphere. Carbon capture, by contrast, involves capturing CO_2 at a point source, such as a stream of industrial waste gas, before it escapes into the atmosphere. Carbon capture is a way to avoid emissions, whereas CDR is a solution to remove emissions that have already accumulated. Much like the separating of recyclable items from trash at home beats the sifting through garbage in a landfill, one can quickly appreciate how carbon capture is more practical than CDR. In addition, capturing carbon emissions before they escape into the atmosphere is both easier and cheaper than using technology to remove emissions that are already in the atmosphere. According to the IPCC, CDR can help resolve these otherwise "hard-to-decarbonize" sectors. It estimates that 100 to 1,000 Gt of CO_2 removal will be necessary during the twenty-first century to ensure that warming goes no higher than 1.5°C.[79] To give some perspective, this will likely require us to be removing up to 15 Gt of CO_2 per year by 2100 – the equivalent of all of humanity's CO_2 output in 1970.

Before proceeding any further, let us not forget that CDR already exists in nature thanks to photosynthesis. Photosynthetic plants and organisms convert carbon dioxide and water into biomaterials, such as sugars and cellulose, using sunlight. In the grand view of the carbon cycle, afforestation strategies, biomass plantations, or algae farming serve the purpose of enhancing the storage capacity of the terrestrial and oceanic sinks. The natural decomposition of biomass does eventually release carbon dioxide back into the atmosphere, through the earth's natural carbon cycle, as discussed in chapter 2. One way of holding the carbon captured by plants indefinitely is to convert it into **biochar** and then bury it deep in the soil to prevent its decomposition back into CO_2.[167] Biochar is a form of charcoal produced by subjecting biomass to very high temperatures in the absence of oxygen. The oxygen-free nature of the combustion prevents carbon from escaping in the form of carbon dioxide, yielding a black brittle material that has

Biochar: a carbon-rich solid residue (charcoal) that remains stable for up to thousands of years; it is often used as a soil enhancer.

an energy content superior to that of its original biomass. Biochar's carbon-rich composition renders it an effective soil enhancer, as discovered by the pre-Columbian peoples of the Amazon who used it extensively over centuries to enrich and make productive what was otherwise sterile land. Its long-term stability makes biochar a particularly effective form of carbon sequestration because it can last in the soil for up to thousands of years.

Biorefinery: a facility, process, or plant that converts biomass into a range of products.

Biofuel: a fuel produced through biological, rather than fossil, sources; biofuels include ethanol, biodiesel, and green diesel.

Alternatively, biomass can be processed in **biorefineries**, facilities that convert biomass into a wide array of products, most notably **biofuels**. Many different types of biorefineries exist at various stages of technological and commercial development, but all are based on the same general concept, that is, breaking down plant matter into its component sugars, starches, oils, and cellulose, using some combination of heat or chemical treatments. Common examples of biomass feedstocks include corn, wheat, cassava, sugarcane, sugar beet, and wood. The versatility of feedstock allows biorefineries to produce a wide range of products, including oil, dietary fibers, pulp and paper, glycerin, and cattle feed. For example, ethanol can be made from corn, biodiesel from palm oil, and aviation fuel from sugarcane.[*] Many important chemicals used in industry can be obtained from lignin, a polymer compound found in the walls of plant cells alongside cellulose that is often a waste by-product of biorefining operations. Recovered lignin can be used for a range of applications, including glues, plastics, and carbon fibers, or broken down further to create a feedstock that can be used to manufacture a wider range of chemicals.

Unlike biochar, the biofuels created from biomass do not offer long-term carbon sequestration: burning biofuels, after all, releases carbon back into the atmosphere. While proponents of biofuels might argue that making fuels from biomass can help to keep fossil carbon in the ground, the climate impact of biofuels is not so simple. Wood and plants are more renewable sources of carbon when compared

---

[*] A 2019 report found that replacing all conventional aviation fuel with sustainable aviation fuels, though theoretically possible, would require some 170 new biorefineries to be built per year from 2020 to 2050.[168]

to fossil reserves; however, this rests on the assumption that the rate of plant regrowth is adequate to offset biofuel's carbon emissions. Dynamic lifecycle analysis of biofuels from wood estimates that it would take between 44 and 104 years for biofuel emissions to be recaptured by the terrestrial sink.[169] So, although it could potentially play a role in managing carbon emissions over the long term, the time scale of biofuel's carbon payback does not align with our short-term mitigation goals. It becomes even harder to justify the use of biofuels when solar energy is undisputedly scalable to any future demand, is already used on a large scale to produce heat and electricity, and poses little or no competition with agricultural land. This said, biofuels could make sense in some specific contexts. For example, a sawmill could be powered by its own wood scrapes, where biomass waste is generated regardless, and jet fuel could be made from biomass instead of exclusively from refined petroleum sources.

Today, however, carbon capture technologies are certainly not limited to natural sequestration mechanisms in the form of biomass. Chemical separation technologies can filter out carbon dioxide from highly concentrated streams of industrial waste emanating from smokestacks. Steel mills, refineries, and power, cement, and fertilizer plants all are suitable carbon capture sites because they offer point sources of carbon. Currently, there are twenty large-scale carbon capture projects worldwide, most of which are integrated with industrial operations. According to Julio Friedmann, senior research scholar at Columbia University's Center for Global Energy Policy, these facilities currently capture 40 million tonnes of CO_2 per year – the equivalent of taking eight or nine million cars off the road.[170] Unsurprisingly, sectors possessing stubborn emissions, namely chemical, iron, steel, aluminum, and cement manufacturing, are the most common to employ carbon capture technologies.

When carbon dioxide is sequestered via biomass, the plant's natural photosynthetic mechanism takes care of the hard work. Capturing carbon dioxide without the help of photosynthesis, it turns

Absorption: a process in which a fluid is dissolved by a liquid or a solid.

Adsorption: a process by which atoms or molecules adhere to a surface.

Cryogenic: occurring at very low temperatures, typically less than −150°C.

Amines: chemical compounds that contain a nitrogen atom and a pair of lone electrons.

Flue gas: gas exiting a pipe into the atmosphere.

Carbamate: an organic compound derived from carbamic acid.

out, is not trivial. Several strategies for capturing CO_2 from both concentrated and diluted sources are currently under investigation. These include physical and chemical **absorption** into solutions, **adsorption** by different classes of solid materials, and **cryogenic** and membrane separation technologies.

The most common method for capturing carbon dioxide at atmospheric pressure from large-scale industrial emitters is by using **amine** solutions in a chemical absorption process. Amines are renowned for their ability to react reversibly with carbon dioxide, meaning that while they can easily react to form chemical bonds with CO_2, under the right conditions, the reverse is also possible. The direction of the reaction is controlled by temperature, allowing CO_2 molecules to be "picked up" and subsequently "released" as needed. In practice, this involves flowing a **flue gas**, typically composed of 10–15 percent CO_2, through two chambers connected by a recirculation system. Both chambers contain an amine solution; however, the first chamber is held at a lower temperature of roughly 40°C, and the second sits at a higher temperature of 100°C. Upon passing through the first chamber, CO_2 molecules in the flue gas react with the amine solution to form **carbamates** and bicarbonates. The flow then continues to the second chamber, where the higher temperature causes the release of relatively pure CO_2 from the carbamate-bicarbonate solution. In practice, the process takes place in two towers, referred as the absorber and the stripper. The gaseous waste stream rises up the absorber tower, where it contacts a downward flow of CO_2-catching amine solution. The CO_2-rich amine solution is then sent to the stripper tower, where heat from rising steam releases, or strips, the CO_2 from the amine, which is returned to the absorber.

Although the amine CO_2-capture systems are well developed, they have several drawbacks including the energy consumption needed for the second chamber, as well as corrosion and degradation limitations of the CO_2-capture absorbents. Research in this area is therefore mainly focused on increasing the rate at which the amines react with CO_2 and decreasing the temperature required for the release step. Improvements can be made through slight modifications to the chemical structure of the amine molecule. The difficulties experienced with amine solutions can alternatively be circumvented by using solid sorbents – typically, amines chemically tethered to silica particles or porous materials – such as zeolites and metal organic frameworks (MOFs), which have high and selective absorption capacities for CO_2.

Sorbent: a molecule used to absorb liquid or gases.

Zeolite: aluminosilicate minerals characterized by their microporous structure.

Metal organic framework (MOF): a class of compounds composed of metal atoms attached to organic molecules that form a characteristically porous structure.

According to the laws of thermodynamics, the higher the concentration of CO_2, the less energy is required to capture it. Simply put, it is easier to collect CO_2 from concentrated sources, such as flue-gas streams, than from thin air. For this reason, carbon capture is most effective, in terms of both energy and cost, when applied at point sources, such as cement manufacturing, chemical refineries, and power plants.[171] Capturing CO_2 from more concentrated sources, however, is not without its challenges. The high rates and variable compositions of flue-gas streams can make efficient extraction of pure CO_2 tricky. Anyone who has used a water-filtration system knows that the process is certainly not instantaneous. Similarly, filtering CO_2 out of industrial flue-gas streams involves a trade-off between the speed of the process and the quality of the product. Separating carbon dioxide from a stream of industrial waste gas can therefore be made easier by ensuring that the stream is as concentrated as possible. Power plants that employ carbon capture technologies can take one of three different approaches to purifying their waste carbon emissions so that they might be more readily captured. These approaches are known as post-combustion, oxy-fuel combustion, and pre-combustion.

Post-combustion is the simplest approach from a process standpoint. It involves sending the untreated stream of flue gas directly through a carbon-dioxide-capture unit, such as the one based on amine-solution technology described previously. No special tweaks are needed to the process: fossil fuel is combusted in air, and carbon dioxide is separated from the waste stream. Post-combustion carbon capture is used at the Boundary Dam Carbon Capture Project in Saskatchewan, Canada, which sequesters nearly one million tonnes of CO$_2$ per year at the site of a coal-fired power plant.

Oxy-fuel combustion and pre-combustion employ further steps to produce more concentrated CO$_2$ streams. In oxy-fuel combustion, oxygen is first separated from air and then mixed with recycled flue gas. The fuel in question is then combusted by this mixture of oxygen and flue gas, rather than by air. This produces a stream of flue gas mostly composed of CO$_2$, which can be readily captured. Although oxy-fuel combustion is practiced in some industries, such as welding and metal cutting, in which a heat source is readily available, it has yet to be adopted for power generation due to the high cost associated with separating oxygen from air. Several pilot-scale projects currently exist, seeking to develop oxy-fuel combustion to sequester carbon emissions from power plants.

Finally, pre-combustion involves taking the fuel in question and exposing it to air at extremely high temperatures and pressures to make **synthesis gas** (a mixture of hydrogen, carbon monoxide, and CO_2; also known as syngas), in a process known as gasification. The syngas then undergoes a chemical reaction to yield a rich mixture of hydrogen and CO_2 (the concentration of CO_2 in this mixture can be as high as 50 percent), making it much easier to subsequently capture the CO_2. This is the capture method employed by the Great Plains Synfuels Plant in North Dakota, which produces methane (natural gas) from coal.

Synthesis gas (or syngas): a mixture of hydrogen and carbon monoxide (and often small amounts of carbon dioxide). Its name derives from its use as a precursor to making synthetic natural gas.

Since it began integrating carbon capture into its process in 2000, the facility has captured three million tonnes of carbon dioxide that would otherwise have been released into the atmosphere. In all these approaches the separated stream of CO_2 can be further purified and compressed for transportation.

Up until now, our discussion of non-photosynthetic means of capturing carbon dioxide has focused exclusively on capture from industrial waste streams. As mentioned earlier, capturing from concentrated sources is much easier, energetically speaking, than capturing from dilute sources. It should therefore come as no surprise that 90 percent of the existing non-photosynthetic CO_2-capture capacity currently in operation is integrated with industrial processes to capture from highly concentrated waste streams.[172] However, we cannot neglect that a significant portion of emissions derive from small distributed sources, such as transportation and buildings. It is therefore also important to develop methods that are suited to smaller applications, such as tailpipes and chimneys.

Aside from natural sequestration via biomass, chemical technologies are emerging that can enable direct air capture of carbon dioxide. Interestingly, the chemistry is reminiscent of a high-speed version of that which occurs in the oceans in the earth's carbon cycle.

The approach involves bringing large quantities of air into contact with hydroxide-rich solutions, which can react with CO_2 to form carbonates. Unfortunately the high solubility of these carbonate products makes it difficult to separate and collect them from the solution. A second step is therefore required in which a reaction between the carbonate and calcium hydroxide forms solid calcium carbonate ($CaCO_3$). The CO_2 can be released from this solid carbonate upon exposure to extremely high temperatures (roughly 700°C).

CO_2 from Thin Air

Although capture of CO_2 directly from the air may be difficult, two companies, Climeworks in Switzerland and Carbon Engineering in Canada, are proving that it is not impossible.

The first facility for capturing CO_2 from thin air using an adsorption-desorption membrane technology opened in Switzerland. The pilot demonstration is integrated with a waste-incineration plant, the heat and electricity of which are used to power the CO_2-capture process. The captured CO_2 is utilized in a large greenhouse located next to the facility to enhance the growth of plants and vegetables. The company responsible for the development and scale-up of the technology is Climeworks, a spin-off from research pioneered at ETH Zurich. Its technology is modular and can be assembled into small, medium, or large capture facilities anywhere in the world. With a current capture capacity of 900 tonnes per year, Climeworks envisions that 250,000 of these direct atmospheric-CO_2-capture plants will be installed around the globe by 2025. The technology has the potential to cut GHG emissions by about 1 percent, which, despite seeming small (roughly the equivalent to the impact of the global fishing fleet),[173] would nonetheless make an important contribution to the global mitigation strategy.

More recently, Carbon Engineering, located in Squamish, British Columbia, is also proving that direct air capture of carbon dioxide is feasible. Its technology is based on hydroxide-carbonate absorption chemistry. Similar to the amine capture process, a large fan is used to suck large volumes of air, which are then subjected to a strong hydroxide solution that captures the CO_2 to form carbonate species. Next, the carbonate solution is reacted with calcium oxide (quicklime) and converted into small pellets of calcium carbonate (limestone), which can then be heated to release pure CO_2. The isolated CO_2 can be sequestered geologically, used for enhanced oil recovery, or converted to chemicals or fuel. The process has the potential to capture up to one million tonnes of CO_2 from the air every year, at a cost of roughly CAD 600 per tonne. Media reports have presented the cost at as low as CAD 100 per tonne; however, this lower price reflects the model in which captured CO_2 is sold to oil companies for enhanced oil recovery, not the true cost of standalone direct air capture.[174]

Between a Rock and a Hard Place

We have mainly focused on the capture part of carbon capture and storage, but the storage aspect is just as critical. Carbon dioxide can be stored in many ways. No matter the approach, it is key that the storage remains reliably stable over long periods of time to minimize the risk of carbon escaping back into the atmosphere.

Geological storage is one of the most common approaches to storing carbon dioxide beyond sequestering it in the form of biomass. As the name suggests, it involves compressing gaseous CO_2 until it

Supercritical state: a phase a of matter in which a substance possesses properties in between those of a gas and those of a liquid.

reaches a **supercritical state**, then injecting it into deep underground reservoirs of porous rock, where it should remain indefinitely.

Maintaining CO_2 in a supercritical state is key to geological sequestration. Although CO_2 is a gas under ambient conditions, it transforms into a supercritical fluid – a state that endows it with properties that are intermediate between those of a gas and those of a liquid – under extremely high pressure.[*] Most important, supercritical CO_2 is roughly three hundred times more dense than gaseous CO_2 (600 kg/m^3 compared to 1.98 kg/m^3), meaning that significantly larger quantities can be sequestered in a given volume of rock. The supercritical state is also desirable for geological storage because, in this form, the CO_2 still expands, much like a gas, to fill the porous structure of the rock. In other words, supercritical CO_2 exhibits the low viscosity reminiscent of a gas, while retaining the high density of a liquid. To remain in its supercritical state, the CO_2 must be subject to extremely high-pressure conditions, which is why it must be stored very deep (typically ~1 km) underground.

Not all rock can securely store CO_2, and storage locations are therefore carefully selected by expert geologists and geological engineers. Gaseous carbon dioxide is injected into a region of underground porous rock that is covered by a layer of impermeable rock. The impermeable layer acts as a seal to contain the carbon dioxide and prevent it from leaking back into the atmosphere. The physical properties of CO_2 also need to be considered. Once stored, it can react with the minerals in the rock to produce carbonate compounds. This process effectively turns carbon dioxide into part of the rock in a process known as mineral trapping. Understanding the chemistry

---

[*] CO_2 becomes a supercritical fluid at 31.1°C and 72.9 atmospheres.

of these processes is critical to ensuring that geological storage remains safe and effective over the long term.

The worldwide capacity for storing CO_2 in underground sedimentary formations is estimated to range from 6,000 to 25,000 Gt.[175] This means that, even with the most conservative estimates, there is more than enough underground storage available to address all the excess CO_2 in the atmosphere.* This is only true, however, if, once stored, the carbon remains stable in the rock formations and does not leak out. So, how long can carbon dioxide be safely stored in geological formations? One study determined the CO_2 retention in reservoirs to be 98 percent over a period of ten thousand years for well-managed reservoirs, and 78 percent for poorly regulated ones.[176] Another way of seeing it is that the risk of leakage from geological reservoirs is considered to be comparable to that of natural gas storage, which is also very low. Still, management of geological storage sites is critical to ensuring their long-term viability, and continuous monitoring is needed to help detect any problems as early as possible.

Another storage option for CO_2 is in the oceanic sink, which has an estimated storage potential of roughly 1,000 Gt. However, this is unlikely to be a wise choice given the potential effects of released CO_2 on marine organisms. Chemical species in solution or liquid pools on the ocean floor risk changing the acidity and ionic constituents of the local environment, which in turn have adverse effects on ocean ecosystems. Large-scale storage of CO_2 in the oceans is also riskier, with the release of carbon dioxide expected to occur over hundreds, rather than thousands, of years. Given the large capacity of geological storage and the already ongoing ocean-acidification problem, it would be wise to leave our oceanic sink alone.

---

* To put this in perspective, recall that we currently emit roughly 33 Gt of CO_2 globally every year.

Carbon capture and storage can not only help to meet our short-term emission targets but also allow us to manage the atmospheric CO$_2$ level and global temperature in the long term. And it is, to some degree, technologically possible. However, if this is all sounds too good to be true, you are right. The fact remains that carbon capture and storage are not considered economically viable at present. Capturing CO$_2$ directly from air, or even from a smokestack, though possible, remains expensive and energy intensive. Profit is usually the overriding consideration when it comes to commercializing a process. Ultimately, large-scale deployment of carbon capture and storage facilities is required for the industry to fulfill its full emission-reduction potential.* The International Energy Agency estimates that over two thousand large-scale facilities would need to be built, requiring hundreds of billions of dollars in investment. However, current policies are insufficient to ensure that industry scales up at the rate required. The main challenge lies in the fact that carbon capture and storage technology is relatively new and therefore deemed riskier by investors. This could be alleviated by putting forth aggressive policy measures that place greater value on reducing carbon emissions. Government can help incentivize the development of carbon capture and storage technology through carbon taxes and grants; some have proved successful in the past.

One of the world's largest demonstrations of carbon capture and storage to date is the Sleipner and Snøhvit project in Norway. Located offshore in the North Sea, near the Norway–United Kingdom border, it captures and stores roughly 0.85 million tonnes of CO$_2$

---

* Although large-scale deployment of CDR and carbon-capture infrastructure is ideal for maximizing their mitigation potential, it is not *required* to render them relevant to climate policy or eligible for investment. In this regard, they differ strongly from geoengineering strategies, which demand an all-or-nothing approach and comprehensive global coordination to be effective. When it comes to CDR and carbon capture, the question of scale need not pose a barrier, because these technologies can be implemented from the bottom up through local initiative, on a case-by-case basis.[177]

every year from a natural-gas power plant and stores it geologically in an offshore deep saline formation. The project came about in response to Norway's 1991 carbon tax, which placed a high penalty on offshore petroleum production that vented carbon dioxide into the atmosphere. Starting at USD 35 per tonne of CO_2 in 1996, it was raised to USD 65 per tonne in 2016.[178] The policy was successful by ensuring that the price on carbon was much higher than the USD 17 per tonne cost of capture and storage.

Still, not all countries have a stringent enough policy on carbon to justify the economics of carbon capture and storage. If capturing from point sources is already non-trivial for companies to justify financially, the situation for direct air capture is worse because the technology is even more expensive. The situation is forcing companies to find alternative ways of ensuring that their carbon capture projects are economically viable, yielding business outcomes that often appear controversial from the point of view of climate action. The business model that has allowed most large-scale capture and storage projects to survive to date has meant working with oil companies to offer their captured CO_2 for **enhanced oil recovery (EOR)**. Of the twenty large-scale carbon capture projects in operation worldwide, thirteen currently sell their captured CO_2 for EOR.

Enhanced oil recovery (EOR): methods used to increase the quantity of crude oil that can be recovered from a reservoir. The most common method involves injecting gaseous CO_2 into the reservoir to reduce the interfacial tension and the viscosity of the oil such that it might be brought to the surface more efficiently.

Enhanced oil recovery involves injecting supercritical CO_2 into oil reserves to help extract more oil from a well. The CO_2 blends with the oil and increases the overall pressure in the reservoir, forcing the oil toward the production wells. The relatively low viscosity of supercritical CO_2 compared to that of other reservoir fluids also allows the oil to flow more easily. These mechanisms allow CO_2 EOR to help recover 60 percent of oil in a reservoir, which is substantial when one considers that standard extraction methods recover no more than 10–40 percent of a reservoir's oil content.[179]

Proponents of EOR justify its environmental benefits by pointing out that it effectively lowers the carbon footprint of oil. After all, roughly 90–95 percent of the injected CO_2 remains trapped in underground rock formation in the space previously occupied by oil. Still, we must not forget the fate of the extracted oil, which releases CO_2 emissions back into the atmosphere upon combustion. If the amount of carbon being trapped underground is greater than that being released by the combustion fuel, can CO_2 EOR be considered a carbon-negative technology?

Estimating the net carbon emission associated with CO_2 EOR projects is unfortunately not straightforward. It depends on several factors, including the source of the CO_2 used, the amount of CO_2 sequestered, and the additional emissions associated with a CO_2 EOR operation – not to mention the emissions that come from the production, refining, and consumption of the resulting petroleum product. This is not helped by the fact that estimating the carbon emissions associated with conventional oil production alone has proved to be non-trivial. The exact emissions are highly dependent on the nature of the individual oil project and may vary greatly between locations. Interestingly, when it comes to CO_2 EOR, studies have shown the potential for decarbonization to be highly time dependent: more CO_2 tends to be released as oil production declines. Specifically, EOR projects tend to yield negative emissions during the first six to eighteen years of operation and then become carbon positive.[180] The timing is therefore critical when one considers the role and impact of CO_2 EOR in the greater climate-action strategy.

One thing, however, is clear: a CO_2 EOR scheme can only offer emission-mitigation benefits if the CO_2 used is captured from an anthropogenic emissions source, such as an industrial waste stream, or is removed directly from the atmosphere. Unfortunately this is not the case at present. Most EOR projects use CO_2 drawn from geological resources in the ground because of the lack of CO_2 located close to oil fields. In the United States there exists over 6,000 km of CO_2 pipeline infrastructure to simply transport CO_2 to the site of EOR operations. Using carbon dioxide from geological sources essentially uses more fossil resources to extract fossil fuels. At best, drawing CO_2 from the ground for CO_2 EOR projects simply displaces geologically stored carbon from one location to another; however, this is challenging given the inevitability of leaks, and some is guaranteed to be lost to the atmosphere along the way. A sobering 70 percent of the CO_2 injected in oil wells today for EOR in the United States comes from geological sources.[172]

Optimists might argue that a lower carbon version of oil-and-gas production is still better than nothing. Integrating carbon capture with CO_2 EOR has the potential to reduce significantly the oil industry's carbon footprint if the carbon used is drawn only from anthropogenic sources. Although the oil industry's current model counts CO_2 as a cost to be minimized, strong policy could compel operations to maximize the CO_2 sequestered in the process. Moreover, CO_2 EOR is currently the only large-scale permanent carbon-sequestration operation that is profitable. Under the right policy measures, it could help jump-start large-scale carbon capture and storage projects by creating a financial incentive to capture and sequester carbon geologically.

At the same time it is not surprising that carbon capture and storage as a mitigation strategy remains controversial among environmentalists who see it as a Band-Aid solution or even an excuse to prolong the era of fossil-based power generation. Carbon capture and storage technologies indeed do fall in a gray area in our current political climate. As Howard Herzog writes in his book *Carbon Capture*, "today the right hates anything to do with climate change, even if it could benefit fossil fuels. Similarly, the left hates anything to do with fossil fuels, even if they can help mitigate climate change. One can say that carbon capture has become an orphan technology."[181]

Whichever perspective you take, it is important to remember some key facts. First, EOR technology can really only be justified if the CO_2 being injected is captured from air or an industrial waste stream; even in this case, it will only offer net negative emissions in the first decade or so of operation. Second, although existing large-scale carbon capture projects tend to have CO_2 EOR operations as their clients for economic reasons, they are not technologically constrained to the oil industry. As we saw earlier, there is no shortage of geological sequestration capacity; storing CO_2 in oil wells is by no means the only option. Carbon capture and storage technologies can exist and operate sans fossil industries and therefore should not be seen as strictly synonymous with the fossil economy. Finally, we cannot turn a blind eye to the evidence suggesting that fossil-fuel companies have been aware of the risk of rising CO_2 emissions for decades, while fostering doubt about the climate crisis.[182] Is it reasonable that these same companies should control large-scale carbon capture and sequestration? Ultimately one must ask whether a decarbonization pathway based on the financial incentive of CO_2 EOR coheres with public interest in the context of a climate emergency.

The intense pressure to reduce emissions drastically in the little time we have left can warrant more Band-Aid-style solutions. In May 2019 Germany, in a struggle to cut its industrial emissions, announced plans to revive a carbon capture and storage project that had been previously halted in 2017 due to mounting pressure from local residents.[183] Although it may not be the ideal choice, challenges in addressing these stubborn emissions can leave leaders with little choice. At the end of the day, some degree of sequestration remains a necessary component to a global carbon-mitigation strategy despite the economic and deployment challenges it faces.

Most important, it should be considered a supplemental tool in the climate-action toolbox, rather than a substitute for conventional mitigation approaches. Just because we have ways of removing some atmospheric carbon dioxide does not justify prolonging emission-intensive industrial practices.

A New Way of Thinking

Richard Buckminster Fuller, author, architect, engineer, inventor, entrepreneur, polymath, and humanitarian, famously wrote: "Pollution is nothing but resources we're not harvesting. We allow them to disperse because we have been ignorant of their value. But if we got into a planning basis, the government could trap pollutants in the stacks and spillages and get back more money than this would cost out of the stockpiled chemistries they'd be collecting." Buckminster Fuller's observation captures the essence of a major paradigm shift that is already underway.

Nobody likes Band-Aid-style solutions, but with stubborn emissions from the industrial sector it might seem like we have no choice. As mentioned earlier, fossil fuels can be replaced by renewable sources for generating heat and energy, but industries like chemical manufacturing use fossil fuels not only for energy but also as feedstock materials. Does a non-fossil feedstock that would fulfill the needs of our chemical manufacturing industry even exist?

Buckminster Fuller's statement was incredibly prescient. As it turns out, the chemistry of the CO_2 molecule lends it to being a key ally in the plight to address industry's stubborn reliance on fossil fuels. Carbon dioxide, the product of burning hydrocarbon fuels, is itself a carrier of carbon and can therefore replace the fossil-derived feedstocks that are conventionally used in industry. Why not use our excess CO_2 to manufacture chemicals and fuels, instead of merely treating it as an inconvenient waste product?

This concept of capturing, recycling, and repurposing CO_2, better known as CO_2 utilization,* is being increasingly endorsed by researchers and policymakers. According to University of Sheffield professor Peter Styring, if 100 percent of **urea**
Urea: an organic (which is used as a fertilizer), 30 percent of min-
compound com- erals, 20 percent of chemicals and polymers, 10
monly used as a
nitrogen source in percent of methane, and 5 percent of diesel and
fertilizers. aviation fuels were made using CO_2, this could consume up to 1.34 Gt every year.[184,185] It is estimated that using CO_2 to manufacture minerals, chemicals, and fuels has the potential to reduce carbon emissions between 10 and 20 percent by 2030.[186] While these numbers are not insignificant, one can see that carbon utilization is certainly not a silver bullet for climate change. Switching our fossil-based energy infrastructure to one that is based on renewable energy and addressing emissions related to land use are still the dominant components of any serious emissions-reduction strategy. Carbon utilization's secret weapon, however, is its potential to eliminate fossil fuels from the supply chain by offering a fossil-free solution to industry's need for a carbon-based feedstock. So, although its emissions-mitigation potential may appear relatively small, carbon dioxide utilization is revelatory because it discredits the view that, even with all the renewable energy in the world, our society cannot survive without fossil fuels.

Carbon dioxide utilization, however, is not simply a scheme to monetize CO_2. The sustainable economy of the future would ideally see the resource-to-product-to-market sequence replaced by a circular system, in which all "waste" product would instead be

---

* Although the word *utilization* has become more or less synonymous with *use*, it holds the specific meaning of using something in a manner for which it was not originally intended. Carbon dioxide utilization implies putting CO_2 to use in unconventional ways, outside its natural role in the earth's carbon cycle, by transforming it into chemicals, fuels, and minerals.

recycled, and resources would no longer be viewed as finite. The world's current manufacturing economy, in the simplest terms, involves the extraction of natural resources (fossil fuels, minerals, lumber, water, etc.), which are then refined and processed by industry to produce valuable commodities (fuels, chemicals, materials, pharmaceuticals, electronics, etc.). The CO_2 emissions that are generated at every step in the process are left to the atmospheric and oceanic reservoirs, where they contribute further to the greenhouse effect and ocean acidification. If, however, we were to treat CO_2 as a useful resource instead of a liability, we could finally get a handle on climate change by closing the carbon loop, as shown in figure 12. It is, however, crucial that in the process of building a future **green economy**,[187] the commodification of atmospheric carbon not be exploited by growth-oriented, free-market thinking for short-term gains.

> **Green economy:** defined by the UN Environment Programme as an economy "that results in improved human well-being and social equity, while significantly reducing environmental risks and ecological scarcities. It is low carbon, resource efficient, and socially inclusive."

The conversion of excess CO_2 into value-added products will play a critical and important role in moving both the transportation sector and the industrial sector into the fossil-free-energy economy over the short and near term. The chemical industry's current supply chain comprises about forty thousand chemicals made from oil, gas, and biomass, granting it enormous potential to adopt CO_2-utilization technologies. Similarly, the conversion of CO_2 to liquid hydrocarbons could hold the key to decarbonizing the transport sector's stubborn emissions associated with large transport vehicles.[188,189]

A global carbon-dioxide initiative is already underway with strategies in place to harness market demand for products that capture and reuse

> **Olefin:** a class of hydrocarbon molecules that contain one or more pairs of carbon atoms linked by a double bond. They are important in the manufacturing of chemicals, plastics, and synthetic rubber.

Figure 12. **The circular CO$_2$ economy**, in which fossil resources are eliminated, and CO$_2$ is continuously recycled to produce valuable commodities.

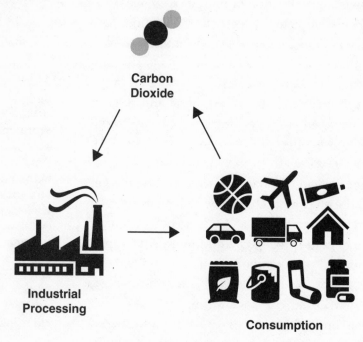

Carbon Dioxide

Industrial Processing

Consumption

Aromatic: a class of hydrocarbon molecules with characteristic planar ring structures. They are used in the manufacturing of a variety of chemicals, including dyes, as well as polymers and synthetic fibers.

CO$_2$.[*] We already have the technical know-how to incorporate CO$_2$ into the industrial supply chain for the production of key chemicals, such as carbon monoxide, **olefins**, **aromatics**, **ammonia**, and **methanol**.

The utilization of CO$_2$ is not an entirely new concept. A number of large-scale industrial processes that consume CO$_2$ were developed during the period from 1880 to 1893. These processes include the synthesis of urea from ammonia;

---

[*] See https://www.globalco2initiative.org/ for full details on the Global CO$_2$ Initiative.

the **Solvay process** for making glass, soap, and paper and for bleaching fabric; the synthesis of sodium hydrogen carbonates, the most common use being baking powder; and the production of salicylic acid, a feedstock for many chemicals, the best known of which is Aspirin.

One need not search far to find ways in which CO_2 already enables many modern technologies. Its applications include soft drinks, dry-ice solid refrigerants, ingredients in frozen foods, enhancers for plant growth and greenhouses, the cooling of bunches of grapes in winemaking, reactive atmospheres for welding, capsules for air guns, extinguishers for electrical and oil fires not put out by water, a supercritical solvent for the removal of caffeine from coffee, polymer processing, chromatography separations, near-infrared gas lasers, ingredients for construction materials such as cement and concrete, fumigants to increase shelf life and remove infestations, and even euthanasia for laboratory research animals.

Methanol: a volatile and colorless liquid consisting of a methyl group bonded to a hydroxyl group. It is widely used in chemical manufacturing.

Ammonia: a compound containing three hydrogen atoms bonded to a nitrogen atom. Its uses include fertilizer, cleaning products, refrigeration, and pharmaceuticals.

Solvay process: an industrial process for making sodium carbonate from calcium carbonate (aka limestone), ammonia, and brine.

The modern chemical manufacturing industry, however, still relies predominantly on fossil-derived feedstock, and until very recently CO_2 has been viewed as a waste product rather than a valuable commodity.[*] The situation today is changing. Mounting pressures to address our rising emissions and the growing demand for energy, food, medicine, and consumer goods around the world are requiring us to seriously rethink our industries. It is rather ironic

---

[*] A slight interest in CO_2 emerged in the 1970s when it was found to be a beneficial additive in the production of methanol, as well as useful for creating improved solvents for lithium-ion battery electrolytes, like propylene carbonate.

that the world's most maligned molecule might just help transition away from a fossil fuel–based economy.

KEY TAKEAWAYS

• Aspects of the chemical manufacturing and transportation sectors possess "stubborn" emissions that cannot be resolved by switching to renewable energy.

• It is easier to capture carbon dioxide from concentrated, rather than dilute sources, such as industrial waste streams.

• Carbon storage is only effective if it can ensure safe, long-term storage, with minimal leaking of carbon dioxide into the atmosphere.

• Carbon dioxide can be safely stored in geological reservoirs without the risk of leakage, provided the sites are well selected and carefully monitored.

• Carbon capture and storage, although technologically feasible, remains costly, and large-scale deployment is not feasible under current policy.

• Enhanced oil recovery (EOR) can help reduce the carbon footprint of oil in some cases, but only if the carbon dioxide used is sourced from the atmosphere or captured from anthropogenic point sources and stored underground post-EOR.

• Carbon capture and storage should be considered a supplemental strategy rather than a substitute for conventional mitigation approaches.

• Fossil-derived feedstocks conventionally used in industry can be replaced with captured carbon dioxide.

- The sustainable economy of the future would ideally see the resource-to-product-to-market sequence replaced by a circular system, in which all "waste" products would instead be recycled, and resources would no longer be viewed as finite.

- Carbon-dioxide removal, capture, storage, and utilization are just some of many key components forming the larger climate-change action strategy to achieve net-zero emissions.

Power to the CO$_2$

We stated earlier that CO$_2$ can replace fossil fuels because it too contains carbon. Chemically speaking, however, fossil resources are hydrocarbons, meaning that they contain both carbon and hydrogen. Carbon dioxide can offer a substitute for the carbon, but a hydrogen source is still needed. Furthermore, many industrial processes still require a source of energy to make them operate. Large-scale deployment of CO$_2$ capture, storage, and utilization, hydrogen production, renewable energy harvesting, and energy-storage technologies are mutually dependent, and their full integration will be necessary for the sustained operation of an emissions-free economy. In this chapter we delve deeper into the technology and auxiliary systems required to turn waste CO$_2$ into chemicals, minerals, and fuels.

To appreciate how CO$_2$-utilization technology works, we must first understand some particulars about the chemical and physical properties of CO$_2$. The CO$_2$ molecule consists of a single carbon atom attached to two oxygen atoms in a linear fashion, as illustrated in figure 13. The central carbon atom shares two electrons with each oxygen atom, which creates very stable bonds. The stability of a molecule is described by a quantity known as its **free energy of formation**. In the case of CO$_2$, its free energy of formation

Free energy of formation: the energy needed to form one mole of a substance in its standard state from its constituent atoms in their standard states.

Figure 13. **The structure of the CO_2 molecule**, consisting of a carbon atom bounded by two oxygen atoms.

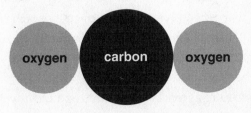

is −394 kilojoules per mole (kJ/mol).[*] The fact that it is negative means that CO_2 has a strong preference to exist as the CO_2 molecule, rather than as its free elements. Conversely, if the free energy of formation were positive, it would indicate that energy would be needed to create the molecule. The counterpart to CO_2 being easy to make is that it is difficult to break. This has consequences for its ability to participate in chemical reactions: the larger the change in **oxidation state** of the carbon in CO_2 on its being converted to product, the more demanding the free energy required to drive the reaction. Simply put, the more the carbon bonds and the electrons that hold them together must be rearranged, the more energy input is needed. The challenge of doing chemistry with CO_2 relates to overcoming its exceptional stability.

Oxidation state: the degree of oxidation, that is, the number of electrons lost from an atom.

Overcoming Barriers

Although CO_2 might be hard to break, it is not impossible; it simply requires enough energy and a little help from **catalysis**. Catalysis is a key concept in the science of CO_2 conversion. There exist several approaches, each defined by

Catalysis: the enabling of a chemical reaction induced by the presence of a catalyst. The catalyst material itself is left unconsumed by the reaction.

---

[*]   The kilojoules per mole refer to the amount of energy needed to create a compound from its constituent atoms.

Activation energy: the energy required for a chemical reaction to occur.

the reaction conditions and type of energy input. Catalysis is enabled by compounds known as catalysts. They lower the **activation energy** barrier of chemical reactions, allowing them to proceed faster and to occur under milder conditions. To elaborate further on what this means, let us return to our cooking analogy from the previous chapter. The reactants in a chemical reaction can be thought of as the ingredients of a recipe, and a recipe provides the instructions on how to transform ingredients (i.e., reactants) into bread (i.e., products). Although an energy input, usually heat, is often required to bake bread, bread does not bake itself. A cook is necessary to gather, prepare, and use the ingredients according to the recipe. From this standpoint, a catalyst can be thought of as the cook who puts the recipe into action. And, much like a good cook, a good catalyst should remain intact after the reaction (be careful of that hot stove!).

Catalysts generally refer to any species that assist in a chemical reaction. Biological catalysts, known as enzymes, play a critical role in the many biochemical processes that keep us alive. Although we will address some examples of enzymes used in CO_2-utilization technologies, most of the catalysts we will talk about are non-biological materials. Currently several different types of catalytic CO_2-conversion processes are under active investigation, all vying for a stake in the race to utilize CO_2 as a supply-chain feedstock. In all cases the discovery of high-performance catalysts is critical to achieving carbon-neutral, efficient, and scalable processes.

Heterogeneous catalysis: a type of catalysis characterized by the catalyst reactants and catalyst having different phases.

All catalytic processes can be classified into one of two main categories: **heterogeneous catalysis**, in which the catalyst and the reactants are made up of different phases (e.g., a solid catalyst reacting with gaseous reactants); or **homogeneous catalysis**, in which they are of the same phase (e.g., both the catalyst and the reactants are dissolved and interact in a liquid).

In the case of homogeneous catalysis, the catalytic and reactant constituents are typically interacting in an **aqueous** or a non-aqueous liquid phase. This results in every catalytic component presenting a unique **active site** to reactants, which often endows them with higher activity and selectivity. Heterogeneous catalysis, however, most commonly involves gaseous or liquid reactants interacting with the surface of solid catalyst material. In this case, the active sites are limited to the surface of the catalyst such that a catalyst with a larger surface area can pack more active sites. The surface area, or the surface-to-volume ratio, is a key factor in determining the activity of a heterogeneous catalyst compared to that of its homogeneous counterparts. A schematic showing the difference between homogeneous and heterogeneous catalytic processes is shown in figure 14.

Homogeneous catalysis: a type of catalysis characterized by the reactants and the catalyst having the same phase.

Aqueous: characterized by the presence of water.

Active site: the part of a catalyst with which reactant molecules interact.

To illustrate better the point of surface area and reactant-catalyst interactions, consider the various ways of cooking potatoes. Mashed potatoes, which involves boiling potatoes until they break down and then mixing them with melted butter, can be thought of as a homogeneous approach to making potatoes because the two components (i.e., the potatoes and the butter) interact in a more or less similar phase. Alternatively we can make roasted potatoes, which involves cutting potatoes into chunks and roasting them in the oven with oil. While the potatoes are baking, the oil is likely to interact only with the surface of the potatoes, resulting in the crisp potato skin that so many of us love. The potatoes being in solid form, and the oil being a liquid, limits the interaction between the two, illustrating the concept of a heterogeneous process. We can appreciate, simply from our knowledge of cooking, that different approaches to preparing the same ingredients can lead to excellent outcomes.

Figure 14. **Schematic illustrating the difference between heterogeneous and homogeneous catalysis**. In heterogeneous catalysis the catalyst is of a different phase (typically solid) compared to the reactant species, whereas in homogeneous catalysis they are of the same phase.

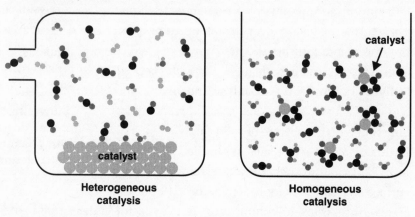

Heterogeneous catalysis

Homogeneous catalysis

Similarly, in catalysis, neither homogeneous nor heterogeneous approaches are necessarily better. The choice lies primarily in catalyst performance, scalability, the resources and infrastructure available for the process, and ultimately cost. This said, for reasons of practicality, heterogeneous catalysis is often favored over homogeneous catalysis because its inherent heterogeneity allows easier recuperation of the catalysts from the reaction medium. Specifically, recovery and reuse of a heterogeneous catalyst is generally more straightforward, less energy intensive, and more cost-effective than that of a homogeneous one that would require a more complicated series of distillations and/or precipitations.

Returning to our kitchen analogy in which the catalyst can be thought of as a chef preparing a recipe, we know that, despite his or her set of skills and experience in the kitchen, a chef alone is not sufficient to cook the ingredients. Cooking food requires a source of energy, such as heat in an oven or radiation from microwaves. Similarly, catalysts enable reactions by lowering the activation barrier, that is, the amount of energy needed for a reaction to take place. In other words, they have the capacity to modify the **kinetics** of a reaction.

Kinetics: the area of physical chemistry concerned with chemical reaction rates.

Consider a mountain pass separating two valleys. A hiker located in one valley wishes to traverse into the neighboring valley; however, to do so, she or he must ascend and descend the mountain separating the two valleys. The height of the mountain can be thought of as the energy barrier that the hiker must overcome to get to the neighboring valley. Getting from one valley to the other is easier if a small hill, rather than a jagged mountain, separates them. Catalysts are conceptually equivalent to small hills: they lessen the energetic barrier that separates those reactants from products in a chemical reaction. Just like hikers (unless they are purposely seeking spectacular views at the highest peaks), reactant molecules will seek the easiest, or lowest-energy, trajectory by which to react.

As previously mentioned, even though the energy requirements are lessened by the presence of a catalyst, most processes still require energy input. Catalytic conversion processes are therefore often characterized by the source of the energy driving the reaction. In the case of electrochemical catalysis, the source of energy is an electric current; for photocatalysis, it is light; and for many processes it is heat. We briefly explore and compare some of the most common catalytic approaches to converting CO_2.

Figure 15 shows the principal types of catalysis that can be used to convert reactants (such as CO_2) into products, classified by their energy source (i.e., heat, light, or electricity). Notice that some types of catalysis draw their energy from two sources, rather than a single source. Catalysis based on photothermal chemistry involves processes that are driven by both light and heat, and catalysis based on photoelectrochemical chemistry uses energy from both light and electricity.

Figure 15. **Different types of catalysis,** classified according to energy source (heat, light, electricity).

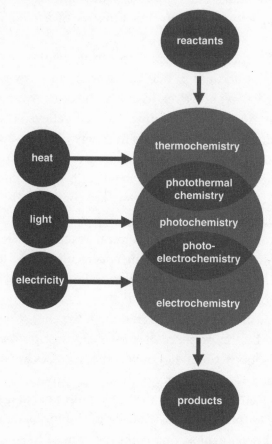

The three energy sources shown in figure 15 can be converted between themselves. For example, light can be converted into electricity, using a photovoltaic device (i.e., a solar cell), or into heat, using a solar thermal device. As well, electricity can be transformed into heat (as in a toaster) and vice versa, using a thermoelectric generator. Although different energy sources are technically interchangeable, doing so is not always favorable, due to reasons of practicality and energy efficiency.

Figure 16. **How all types of catalysis can be driven by sunlight.**

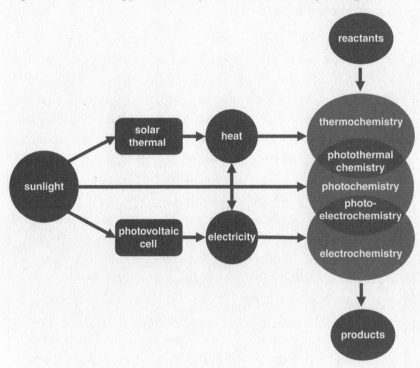

The whole purpose of these processes is to convert CO_2 into useful products in a sustainable manner without employing fossil fuels. Ideally these reactions should therefore be powered by clean sources of energy. It would seem counterproductive to burn coal in order to supply the heat to drive these reactions, even if they are consuming CO_2 emissions. The most abundant source of energy on earth is provided by the sun; enough solar energy strikes the earth's surface in a single hour to fulfill all human needs for a year. To ensure that CO_2 utilization processes are truly sustainable, it would be ideal if all types could be driven by renewable energy, such as sunlight, as shown in figure 16.

A Brief History of Catalysis

Modern society would not be possible without catalysts. Catalysis is needed to manufacture virtually all industrial chemicals and products. Catalyzed chemical reactions are responsible for driving important biochemical processes in all complex organisms. Although catalysis can be traced to the rise of civilization, starting with the fermentation of grains and fruit to make alcohols, it rose to prominence with the seminal work of Paul Sabatier in 1890–1920 on hydrogenation catalysis, which laid the groundwork for the evolution of both heterogeneous and homogeneous catalysis.

A catalyst is a substance that accelerates the rate of a chemical reaction by reducing its activation energy without being consumed in the process. Jöns Jacob Berzelius first introduced the term *catalysis* in 1835. He defined it as the action of a compound that increases the rate of a chemical reaction but does not change during the reaction. Today we know a catalyst *does* change when it catalyzes a reaction, but then it changes back to its original form. Wilhelm Ostwald in 1902 gave a better description of a catalyst as a compound that changes the rate but not the chemical equilibrium; in other words, it modifies the kinetics but not the thermodynamics of a reaction.

It was not until the twentieth century that a deeper understanding of the principles of catalysis began to emerge. This period marked the beginning of the industrial-scale production of bulk chemicals, notably ammonia for fertilizers and explosives, hydrocarbons for fuels, and methanol for manufacturing numerous feedstock chemicals. Although the need for explosives receded toward the end of the First World War, the onset of the Second World War renewed the demand for synthetic fuels and chemicals, which bolstered the development of the chemical industry. When the war ended, many new materials and catalytic processes were being discovered that found applications across many industries. It was around this time that the petrochemical, pharmaceutical, and automobile industries began employing catalysts for the high-volume production of chemicals and fuels. Since then the chemical industry has seen the rise in the use of enzymatic biocatalysis, for example, in the biofuel, detergent, food, brewing, and paper industries.

Today catalysis is indispensable for the large-volume manufacturing of chemicals, polymers, and petrochemicals, with feedstocks deriving predominantly from fossil resources. About 95 percent of these products are made by catalyzed reactions, the majority of which are heterogeneous in nature. The remaining are homogeneous and operate in a single phase, most often liquid, where the catalyst is soluble in the liquid phase. The demand for catalysts today is exceptionally strong and is expected to keep on rising. With our current knowledge of CO_2 catalysis, we can begin to envision the chemical industry of the future in which CO_2 is treated not as a waste product but as a valuable chemical feedstock in a carbon-neutral catalytic cycle.

The Question of Hydrogen

Despite this complete view of the different types of catalytic processes that can convert CO_2 into useful chemicals and fuels, there is a missing piece to the puzzle that we have not addressed: the need for hydrogen gas.* As mentioned earlier, if we want to replace fossil hydrocarbons, we need to find alternative supplies of not only carbon but also hydrogen. Hydrogen is key to realizing CO_2 refineries and is a missing step to completing a fully carbon-negative or -neutral photocatalytic CO_2 process. In fact, the issue is so important that we must pause the discussion on CO_2 to focus on our second favorite molecule, hydrogen.

---

* The term *hydrogen* can refer to either elemental hydrogen or hydrogen gas. In this chapter we will use the terms *hydrogen* and *hydrogen gas* interchangeably.

Haber-Bosch process: an industrial process for converting nitrogen and hydrogen into ammonia, which is used in the production of fertilizer.

Hydrogen is the lightest molecule in existence and is one of the most important chemical products in our society today. Approximately 50–60 Mt of hydrogen are produced annually in the world, and approximately half of this is used in the production of ammonia via the **Haber-Bosch process**, for use as a fertilizer. Another 35–40 percent is used in the petrochemical industry to refine crude oil through hydrodesulphurization and hydrocracking processes. The remaining hydrogen is used in smaller but necessarily important applications such as the hydrogenation of fats and oils; the production of chemicals and petrochemicals, such as methanol, ethanol, and dimethyl ether; the refining of metals and steels; food preparation; rocket fuels; and, more recently, the conversion of CO_2 to useful products. In addition to its use as a chemical, hydrogen can be used as a carbon-neutral fuel: upon combustion, it produces nothing but water or water vapor.

The idea of a hydrogen economy has been a global vision since the oil crisis in the 1970s, and although **fuel cells** were

Fuel cell: an electrochemical device that converts fuel, typically hydrogen, into electricity.

being developed as early as the 1960s, it is only in the last ten to twenty years that the technology has evolved to become a commercial reality. Most major automotive companies, including Hyundai, Toyota, Honda, and Mercedes-Benz, now produce fuel-cell electric vehicles (FCEVs) for commercial use, and many countries have committed to putting more FCEVs on the road in the next five years. Fuel-cell technology is integrated in the vehicle and used in its operation. In this process hydrogen and oxygen produce electricity and water vapor to provide an entirely fossil-free vehicle. Much like electric vehicles, however, they face infrastructural challenges, such as the need to provide refueling stations.

Aside from FCEVs, hydrogen fuel-cell technology also holds much promise for the shipping industry. Recently, ABB and Ballard

Power Systems agreed to work together on the development of a fuel-cell power system for the shipping industry, based on existing kilowatt-scale fuel-cell technologies.[190] The new fuel-cell system is anticipated to constitute a single module that would occupy a space no larger than a conventional marine engine that operates on fossil fuel. Such a technology could have a tremendous impact on the adaptation of the shipping industry to the renewable energy economy.

Although hydrogen may appear to be a strong candidate to replace fossil fuels as the dominant energy carrier of the future, there are issues with its sourcing and cost. Indeed, a sobering 96 percent of the world's supply of hydrogen currently comes from fossil fuels. Steam methane reforming is the most common way of obtaining hydrogen and accounts for nearly half of its production. As the name suggests, the reaction involves applying extremely high temperature to a mixture of methane (natural gas) and water to produce hydrogen and carbon dioxide. The remaining half of hydrogen production comes from petroleum sources and coal. As a result, a large amount of CO_2 is released during its production, with annual emissions topping 500 Mt of CO_2. Clearly, there is an urgent need to develop renewable methods of producing hydrogen if we are to curtail CO_2 emissions.

There are, fortunately, more-sustainable ways of obtaining hydrogen. Water, which is the most abundant and accessible source of hydrogen on the planet, has been known to produce hydrogen through water electrolysis since 1789. Water electrolysis, which accounts for the remaining 4 percent of today's global supply of hydrogen, is achieved by passing electrical current through an aqueous solution and producing hydrogen and oxygen gases. If renewable forms of electricity are used to provide the electrical current necessary to split the water molecules, the process can have a net-zero CO_2 emission profile.

Industrial-scale water electrolysis using alkaline electrolyzers operates at roughly 75 percent efficiencies, which are similar to those associated with steam methane reformation; however, when accounting for efficiency losses that incur from the compression and transportation of hydrogen, as well as during fuel-cell or gas-turbine operation, the overall efficiency drops to roughly 25 percent. This makes it difficult for hydrogen to compete as a fuel source. In addition, the cost of hydrogen production from electrolysis is six to seven times higher than production from fossil-fuel sources. A large amount of this cost comes from the capital cost of renewable sources of electricity, as well as the cost of electricity from the grid, which can vary throughout the day. The energy required to maintain water electrolysis over extended periods of time can be substantial.

Despite the issues with efficiency and energy losses due to the compression and storage of hydrogen, the principal obstacle to wide-scale electrolytic hydrogen production remains an abundant and cheap source of renewable electricity. With the growth in market share of renewable energy, however, we can hopefully expect that the cost of electricity generated via solar, wind, and tidal technologies will decrease over time.

Direct water electrolysis is not the only alternative means of obtaining hydrogen gas. The copper-chlorine cycle, for example, is a three-step cyclic process for converting water into hydrogen and oxygen. In the first step, copper-chloride salt is reacted with hydrochloric acid to give off hydrogen gas, leaving behind a copper-chloride solution. Next, the copper-chloride solution is reacted with water vapor to produce a solid copper-oxide chloride. Finally, the solid copper-oxide chloride is heated, which transforms it back to the original copper-chloride salt, releasing oxygen in the process. The overall process results in the decomposition of water into hydrogen and oxygen gas, and all the other chemicals and materials involved are recycled. The key to the copper-chlorine cycle is achieving the optimal temperature required at each step. In practice, waste heat from industrial processes and nuclear reactors can be used to power the cycle, thereby reducing its operational carbon footprint. Although the copper-chlorine cycle is cumbersome because it requires three separate steps and substantial equipment, it remains a favored method of obtaining hydrogen due to its relatively low energy requirements, its efficiency, and its potential scalability.

Another possibility for hydrogen production involves the use of a catalyst material to make hydrogen from water and sunlight. Photons from sunlight can transfer their energy to electrons in the catalyst, thereby "exciting" them. In their excited state the electrons can be transferred to the water, causing it to split to form hydrogen and oxygen gas. Metal oxide catalysts are ideal candidates for the job because they tend to be good light absorbers. A second catalyst can be used as the site for the water-splitting reaction. The process is chemically equivalent to the reaction that happens in electrolysis, except in this case the electrons are being drawn from the catalyst by light, rather than by an external electric current.

No matter which water-splitting technique is used, the future of sustainable hydrogen production is intrinsically tied to the future of renewable electricity. In the context of seeking climate-change action strategies that are both environmentally sound and socially equitable, the question of water merits some serious discussion. Every continent is already experiencing freshwater shortages driven by a complexity of issues including population growth, pollution, politics, and climate change itself. The number of people without access to a clean and secure freshwater supply is estimated to be nearly a quarter of the world's population. Roughly another quarter live in communities that lack the necessary infrastructure to draw water safely from rivers and aquifers. So, while water might appear to be a more carbon-friendly input compared to fossil-sourced hydrogen, one cannot neglect the fact that, as water becomes an increasingly lacking resource in certain parts of the world, its use for energy generation rather than for drinking is ethically debatable.

**Brackish water:
water containing
salinity levels in
between fresh wa-
ter and salt water,
often resulting
from their mixing in
estuaries.**

Desalination of seawater and **brackish water** seems to be the only feasible way to address
water insecurity and has become an increasingly
important component of the global water supply.
Between 2010 and 2016 the number of desalination installations increased by 9 percent annually.[191] Currently, the global desalination capacity
is estimated to be 95 billion liters per day and is
projected to rise to 170 billion liters per day by 2050.[192] Out of neces-
sity, countries in the Middle East have dominated the desalination
market. However, the threat of freshwater shortages is increasing
around the world, and many countries have opted to install desalination facilities.[193] These desalination processes are often energy
intensive and costly, not to mention often powered by electricity
generated from fossil fuels.

Desalination by **reverse osmosis** and **thermal distillation** are the
two most practiced approaches to obtaining potable water from seawater and brackish water today. In reverse osmosis, water is passed
through a semipermeable membrane containing
nano-sized pores to remove unwanted molecules
and particles. High pressures, created using a
pump, drive the water through the membrane.
Reverse osmosis is the reverse of the principle of
natural osmosis, in which a difference in concentration of ions on either side of a membrane results
in flow from the area of high concentration to that
of low concentration. Thermal distillation involves
evaporating either salt water or brackish water and
then recondensing it to obtain a pure product. Both
desalination technologies require large amounts
of energy. In the case of reverse osmosis, energy is needed to pump
water through the membrane, and in thermal distillation, energy is
needed to produce heat. In both cases, however, renewable energy

**Reverse osmosis: a
water-purification
technique in which
water is filtered
through a semiper-
meable membrane.**

**Thermal distillation:
a water-purification
technology in
which energy is
used to evaporate
water and then re-
condense it.**

can be used to power desalination processes in lieu of fossil fuels. In fact, the use of renewable energy for desalination has increased from 2 percent in 1998 to 23 percent in 2016.[194] Direct absorption of heat from the sunlight can also be used to drive thermal distillation. No matter the configuration, the challenge in all cases is to improve the energy efficiency of the process: in other words, the amount of potable water that can be obtained for a given amount of energy input.

Technological advancements can help to improve the efficiencies of these processes. For example, in the case of solar-powered thermal distillation, mirrors can be used to concentrate the sunlight in order to promote heat generation right at the interface between air and water. The solar-powered thermal water-heating systems used today for domestic and industry purposes have a long history that can be traced back to 214 BC when Archimedes used mirrors to heat water. But there are ways of concentrating light other than mirrors, which is a cumbersome solution. Another way of efficiently harnessing the heat from sunlight is by using a thin membrane that floats on the surface of the water to facilitate the generation of heat from the sunlight. The membrane consists of a porous hydrophobic material that also displays a strong **photothermal effect**. The conversion of solar energy to heat in the photothermal membrane causes local heating at the air-water interface. This in turn causes evaporation of the water through the pores of the membrane, whereupon it can be condensed in a cooler region of the desalination system. The key challenge lies in designing a membrane that is simultaneously porous, stable enough to handle the harsh sunlight and salty water conditions, and sufficiently photothermal in nature. Anyone who has made the mistake of wearing a black cotton shirt on a hot summer's day knows that black materials tend to display a strong photothermal effect. The choice of photothermal materials used in desalination membranes follows this same principle, with typical candidates including

Photothermal effect: the generation of heat by sunlight.

copper, aluminum, steel, and silicon and often having different kinds of textured surfaces to optimize the absorption of sunlight.

Recently a thermal distillation system that used as a photothermal membrane an **electropolymerized** black coating of **polypyrrole** grown on a stainless-steel mesh was reported to produce fresh water successfully from salt water in a 100 percent solar-driven system.[195] The desalination cycle was completed by transferring the evaporated water to a condensing chamber by a solar-powered fan, as depicted in figure 17. A 2019 paper has also demonstrated a means of overcoming the challenges of photothermal membranes by devising a photothermal fabric that hangs over, rather than floats on, the water surface.[196] In the design the membrane hangs just over the water such that only two of its edges are immersed, which reduces heat loss from the system and prevents damage to the membrane caused by salt water.

Electropolymerization: a coating technique in which a polymer is deposited onto a conducting substrate by means of an electric current.

Pyrrole: a ring-structured organic compound. A polypyrrole consists of a chain of pyrroles.

Emerging research in the field of photothermal materials will hopefully help push the development of solar-powered water desalination to meet the growing global demand for safe freshwater sources.

While water is arguably the most renowned means to obtain low-carbon hydrogen, it is not the only one, however. Another option, for example, is to incorporate carbon capture technology with conventional steam methane reforming – in other words, to obtain hydrogen from methane but capture the CO_2 generated in the process to prevent it from escaping to the atmosphere. If the captured CO_2 is sequestered permanently, say in a geological formation, the process becomes a low-carbon route to hydrogen production. Another approach is **pyrolysis**,

Pyrolysis: a decomposition reaction that involves subjecting a material to extremely high temperatures in an inert gas.

Figure 17. **Illustration of an all-in-one solar distillation system for producing fresh water from salt water.**

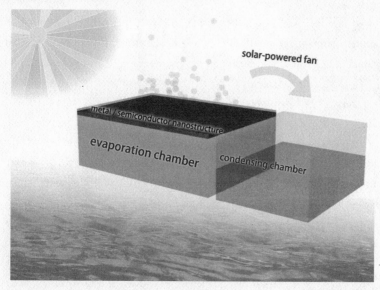

metal / semiconductor nanostructure

evaporation chamber

condensing chamber

solar-powered fan

Courtesy of Dr. Chenxi Qian.

which involves obtaining hydrogen gas by heating methane to extremely elevated temperatures in the absence of oxygen. This results in the formation of solid carbon, rather than the release of gaseous CO_2 that occurs in steam methane reforming. Both approaches have the obvious advantage of not requiring fresh water; however, they fundamentally involve methane, which at present is still predominantly sourced from fossil natural gas. The most sustainable energy economy of the future would ideally be free of all fossil resources, and therefore it makes sense that we should wean ourselves from natural gas as a source of methane. In this regard, biomass feedstock offers a promising alternative. Hydrogen production from pyrolysis of biomass has been found to be a cost-effective route that is competitive with photovoltaic-driven

CO_2 Desalination

Most water-desalination techniques today require large amounts of energy, often supplied by fossil fuels. Could there instead be a way of producing potable water that safely sequesters CO_2, rather than emits it? The latest research shows that this may indeed be possible by exploiting the science behind **clathrate hydrates**. Clathrate hydrates are three-dimensional, cage-like structures made of water molecules; they are essentially tiny ice cages. What is most interesting, however, is their ability to "trap" gas molecules, such as methane, nitrogen, oxygen, and even CO_2. Clathrate hydrates containing CO_2 display pressure-induced **phase transitions** that endow them with unique physical properties relevant in the study of everything from geohazards on earth to the icy bodies in the solar system. They are even being investigated as a means of storing GHGs by injecting them into deep saline aquifers.[197]

Clathrate hydrate: a crystalline solid composed of hydrogen-bonded water molecules that take on a cage-like structure.

Phase transition: the transition of a substance from one physical state or structure to another.

Clathrate hydrates that contain CO_2 tend to crystallize under high pressure and low temperatures in salty water. This feature could open the door to water-desalination applications. In their solid, crystallized state, CO_2-containing clathrate hydrates could be readily separated from salt water. Subjecting them to ambient pressure and temperature conditions would allow the CO_2 to be captured and stored, leaving behind pure H_2O.

Whether or not a CO_2-desalination process of this kind could be commercially viable will depend on numerous factors, especially on the quality of the water produced. Ongoing research and development will hopefully continue to improve the quality and economics of CO_2-desalination to the point that it becomes competitive with membrane-based reverse osmosis and thermal distillation.

water electrolysis. Of course, as discussed earlier, the use of bio-mass as a fuel source is also debatable because the scaling of bio-mass production risks competing with fertile land that would otherwise be used for food production. At the end of the day there is no "free" source of hydrogen; it must eventually come from somewhere, be it fossil fuel, biomass, or water.

Does CO_2 Need Electricity, or Does Electricity Need CO_2?

Throughout this book is the recurring theme of the need for elec-tricity generated from renewable sources. The catalytic conver-sion of CO_2 to chemicals, the electrolysis of water to produce hydrogen, and water-desalination technology all require a car-bon-neutral source of energy. Although some of these processes have the potential to be powered by sunlight directly, they of-ten still require some degree of electricity. For example, in the case of photocatalysis, electricity is needed to supply hydrogen for the reaction.[†] The limiting factor to making most of the pro-cesses economically viable is the cost of carbon-neutral electricity. Developing and expanding the renewable energy infrastructure is therefore crucial to the success of CO_2-utilization technol-ogies, not to mention an imperative step toward decarboniz-ing the world's energy and transportation sectors. Switching to

---

* Fresh water used in agriculture should have a total dissolved-solids level of around 500 ppm, whereas acceptable levels for residential consumption are less than 100 ppm.

† Emerging research is making a case for the possibility of hydrogen production from direct sunlight; however, this remains the exception, and, at present, the electrolysis of water is still the most reliable method of making carbon-neutral hydrogen at scale.

renewable electricity as a primary energy source, however, presents its own set of challenges.

The most commonly raised problem with renewable energy is that of intermittency. Unlike fossil-fueled power plants, photovoltaic cells and wind turbines cannot guarantee a consistent supply to the grid. The answer to this intermittency issue is storage; holding excess energy for later use can ensure a robust and reliable power supply at all times. Here, the use of batteries for large-scale storage applications offers a promising solution by storing excess electricity during off-peak hours and releasing power during peak demand.

More than two centuries after its invention by Alessandro Volta in 1800, the battery remains an indispensable part of modern life. A battery is an electrochemical device that converts stored chemical energy into electrical energy. In its simplest form a battery comprises two electrodes (a cathode and an anode) and an electrolyte contained between. The battery operates by the movement of ions between the electrodes, which produces a concomitant flow of charge out of the battery to an external electrical circuit. Many types of batteries exist today, each lending itself to different energy-storage applications – such as lithium-ion batteries found in portable electronics and electric vehicles, and lead-acid batteries used in cars, buses, electric bikes, and industry. **Redox flow batteries** are one of the most promising candidates for grid-scale storage applications. Their main advantage lies in their power and energy capacity being separately tunable, as well as in their long life cycle. Still, they remain limited by their tendency to discharge and their significant capital costs. Recently lithium-ion batteries, once thought to be limited to vehicles and portable electronics, have also been demonstrated as a viable solution to grid-scale energy storage. The world's largest lithium-ion energy

Redox flow battery: an industrial-scale battery in which charge is stored in the form of liquid electrolytes, allowing for easily scalable storage capacity.

storage facility, located in Australia, uses Tesla's Powerpack batteries in conjunction with a wind farm to power tens of thousands of homes with renewable electricity. Tesla's latest battery storage product, the Megapack, is designed specifically for utility-scale projects. The company has already installed more than one gigawatt hour of electrical storage capacity using its lithium-ion-based battery products.

Although certain types of battery and energy-storage systems are technologically mature, many others are being actively explored, developed, and optimized. For example, in January of 2018, researchers at MIT revealed a new design of a rechargeable liquid-sodium battery that could be used for grid-scale storage.[198] Although the battery itself is not new (it was designed in 1968 but quickly abandoned for reasons of impracticality), researchers discovered new potential when they found that integrating a novel type of metal mesh membrane could render it robust enough for use in utility-scale applications that require low-cost storage solutions.

Until grid-scale battery storage becomes an established component of the electricity infrastructure, the jury is still out on what to do with excess electricity. A contemporary view is that excess electricity could be stored in the form of chemical bonds. For example, this could involve using the electricity to convert CO_2 to chemicals via electrochemical catalysis or to convert water to H_2 via electrolysis. Alternatively, the electricity could be converted to heat and then be used to convert CO_2 to chemicals via thermal catalysis. Whether it is more efficient to use the excess electricity to power CO_2-conversion technologies (either directly or via heat) or to produce H_2 remains unclear at this point. In either case, grid-scale electrical-to-chemical energy storage can be employed to control the load leveling of power and demand of the electrical grid, as well as the production of value-added chemicals and fuels.

Compared to flow batteries, chemical storage of electricity via catalysis and electrolysis could allow for seasonal storage benefits.

This may be particularly useful to electricity-rich provinces, such as Quebec and Ontario in Canada, which could use their excess electricity for production of chemicals and fuels in the summer, which in turn could be used for consumption in the winter. Storing electricity in the form of chemicals and fuels is also the better solution from the perspective of powering large transport vehicles, such as tankers, cruise ships, aeroplanes, and large trucks that still require fuel. Other advantages include the longer storage times, easy transportation across long distances, and the higher energy density in chemicals and fuels compared to batteries.

A counterargument is that there is simply not enough excess electricity at present to render chemical storage economically significant. Instead, the surplus is more wisely used elsewhere, such as in the charging stations that power the rapidly growing number of EVs. Furthermore, batteries are a well-developed and mature technology, and storing grid electricity in batteries and charging units remains more efficient from an energy point of view compared to converting electricity into chemicals and fuels for later use.

The best option for energy storage is highly dependent on a region's geography, existing infrastructure, and local industry. Beyond batteries and catalysis, transmission over long distances offers another solution for addressing intermittency, while also bringing renewable energy resources to remote communities. In some countries, the most energy-efficient and cost-effective way of storing megawatt quantities of excess electricity is in the form of potential energy by pumping water up to mountain reservoirs. In other cases, it can be stored as thermal and mechanical energy by compressing air, or as rotational energy in spinning flywheels. Most recently, the Swiss company Energy Vault has demonstrated a groundbreaking gravitational-energy-storage technology. The process involves using cranes to pick up and stack concrete blocks: during periods of excess solar or wind power, the electricity-driven cranes lift the cement blocks and stack them such that they are higher off the ground, eventually creating a tower of blocks. During periods of intermittency the blocks are lowered by the force of gravity and, in the process, power a generator. The system can potentially store 20 MWh of energy – enough to power over a thousand homes for a day. While such examples can certainly inspire one to reflect on the potential of diverse technologies, we should point out that these electricity-storage methods combined do not yet offer the scale of storage capacity needed. Still, they challenge conventional ideas about what energy storage can look like.

When it comes to imagining the renewable energy infrastructure of the future, the single most important change to consider is the anticipated increase in load on the grid. With previously fossil-powered technologies being replaced by electricity-powered ones, we can only expect the demand for electricity to rise dramatically. A recent paper estimated that carbon capture and utilization technologies would require upwards of 18.1 petawatt hours of low carbon electricity.[199] Grids will need to be upgraded, managed, and maintained to accommodate the large increases in consumption that are to be expected by new sources drawing power. A recent study in Germany revealed that grid-relieving measures, such as major grid expansion and the integration of flexible loads, would be necessary to accommodate the larger rates of consumption.[200,*]

We began this chapter by explaining the various ways to power CO_2-utilization technologies. Just like cooking food, making products from CO_2 and H_2 via catalysis takes energy. Whether this energy takes the form of heat, electricity, or direct sunlight, the whole process should ultimately remain carbon-neutral and be free of fossil fuels. The same goes for the production of H_2, which requires the development of an extensive water-desalination and electrolysis infrastructure. The

---

* The case study analysis estimated that in an aggressive electrification strategy for Germany, electricity-consumption rates in 2030 would be 40 percent higher than those in 2015.

renewable energy needed to power all these processes demands that we rethink our electricity grids and address the issue of intermittency. The point is that, far from being stand-alone solutions, CO_2 utilization, H_2 production, water desalination, renewable-resource harvesting, and energy storage are all mutually dependent systems whose intentional integration will be necessary for the creation of a sustainable, emission-free energy economy. And so, does CO_2 need electricity, or does electricity need CO_2? We leave it to you to decide.

Carbon Sequestration Using Rechargeable Batteries

Sony and Asahi Kasei produced the first commercial lithium-ion battery in 1991. First used in consumer products, these batteries have since evolved to power personal electronics and EVs, as well as larger-format cells in energy-storage applications. Today electrochemical energy storage continues to be an intensive area of research in both academia and industry, drawing from many fields of expertise including electrical engineering, chemistry, and materials science.

Lithium-ion batteries, depending on the required specifications, embrace a wide range of different chemistries. However, they all operate on the same basic principle that involves the transfer of lithium ions between the anode electrode and the cathode electrode in charging and discharging reactions. On discharge, the lithium ions generated at the anode diffuse through a lithium-ion polymer or inorganic solid electrolyte to the cathode into which they insert and/or react. To complete the process, charge-balancing equivalents of electrons transfer from the anode to the cathode through an external circuit, thus generating a current. The discharge reaction is effectively the conversion of chemical energy into electrical energy. By applying an external voltage bias, the process can be reversed, and the battery can be recharged.

In the quest to find ways of sequestering excess CO_2, it is interesting to contemplate how it might be incorporated into this electrochemical operation. The cathode electrode in a lithium-ion battery is typically made of a metal or metal oxide material. We know that these also tend to make good catalysts for converting CO_2 to chemicals. Could the electrochemical reaction in a lithium-ion battery be somehow orchestrated to involve CO_2? Emerging research shows that it can be done using what is known as a Li-CO_2 battery.[201] It involves adding CO_2 to the device such that, upon discharge, the lithium ions react with CO_2 at the metal oxide cathode to form lithium carbonate and carbon. Upon charging, the solid lithium carbonate reverts to lithium ions and CO_2 but the carbon, being very stable, remains fixed. In other words, every time the cell discharges and charges, some carbon is left behind on the cathode instead of being converted back into CO_2. If the CO_2 is continuously replenished, the device can effectively charge and discharge and simultaneously sequester CO_2 in the form of carbon. Although the research is still in the proof-of-concept stage, the Li-CO_2 battery provides an interesting paradigm from which to develop large-scale electrochemical systems not only to store renewable energy but also to sequester CO_2.

KEY TAKEAWAYS

- The challenge of converting CO_2 to chemicals and fuels lies in overcoming the high chemical stability of CO_2.

- Catalysts help to accelerate the rate of chemical reactions and allow them to occur under milder conditions (requiring less energy).

- Converting CO_2 to chemicals and fuels requires energy (i.e., heat, electricity, or sunlight) and hydrogen (H_2). Renewable energy and fossil-free H_2 must be used to guarantee carbon neutrality.

- The most common way of producing H_2 today is from fossil fuel via steam methane reforming; however, it can be produced from the electrolysis of (pure) water, using renewable energy.

- Water scarcity is growing worldwide; water-desalination techniques, such as reverse osmosis and thermal distillation, can help to address the demand for potable water.

- The cost of renewably generated electricity is a limiting economic factor in the catalytic conversion of CO_2 to chemicals, the electrolysis of water to produce H_2, and water-desalination technology.

- Developing and expanding the renewable energy infrastructure is critical to the success of carbon-neutral technologies.

- Batteries can help to address renewable energy's intermittency problem by storing excess electricity during off-peak hours and releasing power during times of peak demand.

- Energy can alternatively be stored in the form of chemical bonds by using excess electricity to convert CO_2 to chemicals, or water to H_2, or sea water to potable water.

- The best option for grid-energy storage is highly dependent on a region's geography, existing infrastructure, and local electricity demand.

- The rise in demand for renewable energy, to be expected as we transition away from fossil fuels, will require major grid expansion and development.

- Large-scale deployment of CO_2 capture, storage, and utilization, H_2 production, water desalination, renewable-resource harvesting, and energy-storage technologies are mutually dependent. Their integration will be necessary to achieve a sustainable, emissions-free economy.

It Is a CO$_2$ World

Collecting facts is important. Knowledge is important. But if you don't have an imagination to use the knowledge, civilization is nowhere.

– Ray Bradbury

Ray Bradbury's statement is especially timely in the context of climate change: what would a world in which humans emit net-zero carbon emissions even look like? Although we might know with certainty that we need to transition away from our current fossil-based economy, how this might look in practice is less clear. Grasping the scale and scope of our current supply chains is already hard enough, let alone conceiving the transformations they will have to undergo. Nonetheless, bold vision will be necessary to lead the transition toward a sustainable and equitable carbon-neutral economy. In this chapter we hope to stimulate your imagination with some of the unexpected and revolutionary ways in which CO$_2$ utilization can help shape the net-zero emission world of the future.

Fossil-Free Fuels

As we saw in chapter 1, oil, natural gas, and coal still constitute our major energy sources and remain the principal chemical inputs

Fischer-Tropsch process: a multi-step chemical reaction involving the conversion of syngas into liquid hydrocarbons.

to a variety of industrial processes. In the current fossil energy infrastructure, fuels are made either directly from petroleum (crude oil and its refined products) or from natural gas and coal. Converting solid coal and natural gas to liquid fuels is done by first reacting them at very high temperatures to form syngas, in a process known as gasification. Next, the syngas is converted to liquid hydrocarbons via the **Fischer-Tropsch process**. A metal catalyst, such as cobalt, iron, or ruthenium, is commonly used to ensure that the reaction produces hydrocarbon chains of the desired length.

It is challenging to conceive of a single chemical that would be able to substitute for a fossil resource's part in the energy, transportation, and industrial sectors. Methanol, however, comes close. In the early 2000s, Nobel laureate George Olah put forth the concept of the "methanol economy," in which methanol and its derivative, dimethyl ether (DME), would replace fossil fuels as a ground transportation fuel and chemical feedstock.[202] It makes a lot of sense: methanol can be used as a feedstock to make synthetic hydrocarbons and their products, as a liquid fuel, and as a convenient energy-storage medium.

Methanol is already used as a feedstock to create roughly 30 percent of all synthetic chemicals. It can also be used as a liquid fuel. Although methanol contains half the energy density of gasoline, it possesses a higher **octane number**, meaning that a methanol-powered engine burns more efficiently compared to a gasoline-powered engine.[*] In addition to offering a better fuel economy, the burning of methanol emits fewer air pollutants, such as hydrocarbons,

Octane number: a measure of the performance of a fuel. The higher the number, the more compression the fuel can withstand before igniting.

---

[*] Less than double the amount of methanol is necessary to achieve the same power output as that of gasoline.

nitrogen oxides, sulfur dioxide, and other particulates, many of which are GHGs. Alternatively, methanol can be converted to DME, which has been touted as a promising fuel for diesel engines and gas turbines. It has an advantage over petroleum-derived diesel fuel because of its high **cetane number**. Another methanol-derived fuel is biodiesel, which is made by reacting methanol with vegetable oils and animal fats. Lastly, methanol can be used in fuel cells, either by first converting it to hydrogen for hydrogen fuel cells, or directly in a methanol fuel cell. Methanol fuel cells are an appealing alternative to hydrogen because of their higher volumetric energy density.

Although methanol has a long and fascinating history, traceable to its use for embalming mummies in ancient Egypt, industrial methanol production only began in 1923. The original process used a chromium and manganese oxide–based catalyst to convert synthesis gas (a mixture of carbon monoxide and H$_2$) into methanol. Since then, improved catalysts have been discovered that allow the reaction to operate just as efficiently but under milder temperatures and pressures. Today the most common catalyst used for methanol production is composed of copper-zinc oxide particles that sit on an aluminum oxide **support**. The copper-zinc oxide catalyst, however, helps further by enabling a more forgiving feed: instead of strictly taking in synthesis gas, the reaction can accommodate some measure of CO$_2$ into the feed, thereby reducing the demand for synthesis gas. Replacing synthesis gas, which is typically derived from fossil sources, with CO$_2$, either from the

Cetane number: a measure of how quickly diesel fuel starts to burn. Diesel engines function differently than do gasoline engines, operating on the principle that the diesel should ignite as soon as it comes out of the injector. A higher cetane number indicates that the diesel fuel will ignite more efficiently, resulting in fewer particulate emissions during combustion.

Support: in the context of catalysis, a material on which the catalyst sits. Although the support itself may exhibit catalytic properties, its main function is to provide a physical structure on which the catalyst can be well dispersed.

atmosphere or waste sources, would be a significant development on the road toward a sustainable methanol economy.

Simply replacing the synthesis gas with CO_2, however, is not without its set of challenges. One barrier to achieving efficient methanol production from a CO_2-based feed is the issue of selectivity. *Selectivity* is a term used by chemists and chemical engineers to describe the propensity of a reaction to create a particular product. For example, a reaction could form product A under a particular set of temperature and pressure conditions, but favor the formation of product B under a different set. In other words, the selectivity of the reaction can change under different conditions, despite the reactants staying the same. In addition to temperature and pressure, the presence of a catalyst can also alter the selectivity of a chemical reaction. In the case of methanol synthesis using a CO_2-based feed rather than a synthesis gas feed, substantial research has been devoted to improving the selectivity of the reaction toward methanol, while using a CO_2-heavy feed and ideally operating under mild temperature conditions. Understanding the role of the copper-zinc oxide catalyst in the reaction, and the ways in which it might be improved, lies at the core of this ongoing challenge. In the meantime, making methanol from CO_2, though not perfect, is already being realized.

Carbon Recycling International, founded in 2006, is a renewable methanol plant in Reykjavik, Iceland, that produces four thousand tonnes of methanol every year. Unlike conventional methanol production, which uses synthesis gas, their process uses H_2 and CO_2, the former of which is made via the electrolysis of water. The electricity used to produce H_2 is generated from Iceland's established geothermal-heating infrastructure, rendering the process completely free of fossil fuels. An international success, Carbon Recycling International's process is a good example of an operation that is mindful of the local market, supply chain, and existing infrastructure. Iceland's vast geothermal resources supply ample energy for the population's heating and electricity; however, the renewably generated electricity is less useful for powering vehicles. The methanol synthesis operation provides a means to transform some of the country's renewable energy resources into a liquid fuel for trucks and cars and consequently to further reduce Iceland's dependence on fossil fuels. According to their website, Carbon Recycling International's renewable methanol produced a mere quarter of the GHG emissions typically generated by standard car fuel. Their solution is also flexible in that the renewably generated methanol can blend with gasoline to be used in flex-fuel vehicles.

The success of Iceland's methanol economy is an example to which regions around the world can aspire. However, applying the concept of the methanol economy on a global scale might not be possible in the short term. Are there other ways of turning CO$_2$ emissions into liquid fuels?

An incredible technological achievement would be the conversion of CO$_2$ directly to gasoline. Perhaps the biggest challenge to realizing this dream is to circumvent the century-old, currently practiced, multistep industrial processing route. By now you might have noticed a trend with many industrial processes involving CO$_2$: they often require multiple steps. For example, H$_2$ needs first to be obtained either from water electrolysis or by steam methane reforming, and then to be reacted with CO$_2$ in an appropriate ratio. If one wishes to produce more sophisticated hydrocarbon products, such as gasoline, further refining steps are needed to increase the carbon content of the product. This three-step formula is cumbersome, particularly with each step requiring its own catalyst. A big breakthrough, however, has come through recently, showing that the gasoline-range hydrocarbons can potentially be produced directly from CO$_2$.[203] The process involves combining three well-known catalysts to form a multifunctional hybrid catalyst. Each component of the hybrid catalyst enables one of the three processing steps, but when chemically integrated, they can enable all three reactions to operate consecutively in a single-step catalytic process. Although still in the early stages of development, this example reveals that the idea of a CO_2-to-liquid-fuel refinery may very well be more than just a dream.

Synthetic fuels from CO_2 and hydrogen or, in some cases, from biomass can offer a sustainable alternative to conventional fossil fuels. Not only can they help to mitigate emissions by utilizing the excess of carbon in our system instead of tapping further into fossil reserves, but also their combustion produces fewer pollutants compared to conventional fossil fuels. Moreover, synthetic fuels can be easily adapted to the existing fuel distribution, storage, and delivery infrastructure. Finally, like most CO_2-based technologies, successful deployment ultimately rests on access to plentiful and cheap renewable energy.

Figuring Out Fertilizer

Global food security depends on the production of nitrogen fertilizers. Without fertilizer, crop yields would be half of the current levels. It is expected that food production will need to increase by 70 percent worldwide over the next thirty years to sustain a growing population.

Urea is a nitrogen-containing fertilizer synthesized from ammonia and CO_2. Some industrial processes already use CO_2 as input, urea synthesis being by far the largest. Fertilizer production is a serious global business. In Canada the fertilizer industry employs over twelve thousand skilled workers, generates CAD 12 billion in annual economic activity, and exports to over eighty countries. In fact, 12 percent of the world's fertilizer is supplied by Canada, and fertilizer is the third-highest-volume commodity shipped by Canadian railways.[204]

Most urea produced today comes from ammonia (NH_3) and CO_2 in what is known as the **Bosch-Meiser process**, which was invented in 1922. The reaction occurs in two stages. In the first stage, CO_2 is combined with NH_3 to form ammonium carbamate. The reaction is highly **exothermic**, meaning that it releases large amounts of heat. In the second stage, ammonium carbamate is subjected to high temperature and pressure conditions, causing it to decompose to urea and water. This second step is highly **endothermic** and is performed using heat generated by the first reaction. The urea product can be retained in the form of an aqueous solution or separated from the water and made into solid granules. In either form, urea can be added directly to soil by using conventional spreading equipment.

Bosch-Meiser process: an industrial process for making urea fertilizer.

Exothermic: characterized by the release of energy.

Endothermic: characterized by the absorption of energy.

It takes roughly 0.7 tonnes of CO_2 to produce approximately one tonne of urea. This is a lot when one considers that a huge amount, 169 million tonnes, of urea was produced in 2015 alone. At first glance, this might appear to be a climate-friendly solution because it consumes lots of CO_2. However, the process operates most efficiently under very high temperatures and pressures, and the large amount of energy needed to create these conditions is typically supplied by burning fossil fuels. In addition, the CO_2 used is most often derived from natural gas – a fossil fuel. As a result, the Bosch-Meiser process possesses a huge carbon footprint, and little has been done to address its environmental impact since its invention almost a century ago.

Bringing urea manufacturing into the twenty-first century requires that we address its CO_2 problem. Research is already being carried out to discover new catalysts and processes that would allow urea to be made in a single step, under much milder conditions. Furthermore, the twelve million tonnes of CO_2 currently produced from natural gas could potentially be replaced by excess CO_2 from the atmosphere. Rethinking the recipe for urea synthesis to address CO_2 emissions need not come at the loss of economic efficiency.

Creating a single-step process that operates under milder conditions would simplify the design of industrial plants and reduce overall operating costs.

There is, however, another, even more important cause of urea's enormous carbon footprint: the ammonia that is needed in the first step of urea's production. Ammonia is produced via the Haber-Bosch process, which like the Bosch-Meiser process, was invented in the early twentieth century. The amount of CO_2 released into the atmosphere from the Haber-Bosch process's supply chain constitutes a startling 3 percent of all GHGs. Moreover, ammonia production accounts for 3–5 percent of the world's natural gas production, or roughly 1–2 percent of the world's yearly energy supply.

The Haber-Bosch process involves mixing nitrogen gas with H_2 under extremely high temperature and pressure conditions of roughly 400°C–500°C and 15–25 megapascals, respectively. The reaction also requires a catalyst, the most common being some form of iron powder. The culprit behind the massive carbon footprint of ammonia production is the H_2 that, as mentioned earlier, is most commonly obtained from steam methane reforming of natural gas. Interestingly, when chemists Fritz Haber and Carl Bosch first began making ammonia, they used H_2 from electrolysis of water. Sourcing carbon-neutral H_2 by electrolysis rather than from natural gas is key to controlling the fertilizer industry's GHG emission problem.

Today scientists are also investigating alternative ways of addressing ammonia's carbon footprint associated with H_2 production.[205] One idea is to make ammonia directly from nitrogen and water, using nothing other than sunlight, dubbed *solar ammonia*. The key to realizing such a technology rests on discovering a photocatalytic material that would allow the reaction to proceed under milder temperatures and pressures. A variety of candidate materials are being studied, but their yields are at present still too low to be considered for industrial applications. Nevertheless, research into solar ammonia is in its infancy, and there is plenty of opportunity for incremental improvements through materials chemistry and reactor engineering. Finding a way to produce ammonia from nothing more than nitrogen, water, and sunlight on an industrial scale could very well make a serious dent in global emissions, while ensuring that the world remains well fed.

Support Space Exploration

Given the pressures to reinvent our fossil-based processes on earth, it might seem counterintuitive to seek solutions in space. However, space-travel technology can provide much inspiration to those of us seeking to manage our resources more wisely on earth. Some very creative space scientists and engineers have been developing innovative ways to sustain the life of astronauts living in the space station. In space, like on earth, the human body needs a constant supply of oxygen and water, and in the confinement of space travel it needs a means of dealing safely with exhaled CO_2 respiration. The average human can only survive about three days without water and suffers permanent brain damage in about three minutes without oxygen.

For astronauts to live and work in space, water, oxygen, CO_2, and H_2 must be orchestrated to provide them with sufficient oxygen and water. This problem cannot be solved by using plants alone, simply because the number that would be required to remove CO_2 from the air is too large and impractical. For this reason a water, oxygen, CO_2, and H_2 closed-loop recycling system is needed for humans to survive in space situations. All the components needed to operate such a recycling system are based on the use of catalysts.

The space station's water–carbon dioxide cycle begins with a hydrogen oxygen fuel cell, which produces water and electricity. The water so formed in the fuel cell passes to an electrochemical system, which regenerates hydrogen and oxygen. The generated oxygen becomes breathable cabin air for the astronauts, while the hydrogen can feed the **Sabatier reaction** to produce water and methane. The

Sabatier reaction: a chemical reaction involving the conversion of hydrogen and carbon dioxide into methane and water.

crew's exhaled CO_2 is captured and is subsequently combined with H_2 in the Sabatier reaction. The water produced in the Sabatier reaction, and water vapor from the cabin air, enter the water recovery system and then go back into the oxygen electrochemical generator system. The methane produced in the Sabatier reaction can be used as a fuel, and the water and the CO_2 emissions can be recycled.

All these processes form a sustainable self-contained system that is able to create a life-supporting atmosphere for the crew. If we can create a zero-waste, closed-loop approach to managing resources in space, the same is surely possible on earth.

Pollution-Free Plastics

If ever there was a controversial global materials success story, it is the story of plastics. Remember the unforgettable statement in the 1967 movie *The Graduate*, in which Benjamin Braddock, an anxious graduate student who is worried about a career, gets some advice from Mr. McGuire: "There's a great future in plastics. Think about it."

Today we can see the prescience in Mr. McGuire's advice. For more than half a century the global production of plastics has grown continuously from around 1.5 million tonnes in 1950 to about 322 million tonnes in 2015, and it continues to grow. At the same time, there is growing concern over the effects of plastics on the environment. According to a recent report put out by the Center for International Environmental Law, if plastic production and use continue along their current path of growth, the carbon footprint of the plastics industry could reach 1.34 Gt per year by 2030, the equivalent of introducing up to three hundred new coal power plants.[206] It is estimated that 15 percent of plastics in Europe and only 9 percent of plastics in the United States are eventually recycled.[207] Current practices associated with the manufacturing and disposal of plastics pose a serious challenge to keeping global temperature rise below 1.5°C. With plastics being so critical to our modern infrastructure, it is critical that we decrease the industry's massive carbon footprint.

Polymers* can come from either a natural source, such as rubber, or a synthetic source, such as ethylene or propylene. Synthetic rubbers, however, are typically derived from petroleum-based fossil fuels. Chemically speaking, polymers are macromolecules with a long-chain structure comprising repeating molecular building blocks. There are many classes of polymers, with wide-ranging architectures and diverse chemical and physical properties. They can be classified as *standard plastics*, such as polyethylene, polypropylene, polystyrene, and polyvinylchloride, or as *engineering plastics*, such as tetraphthalate and polycarbonate.

Polycarbonates are a class of engineering plastics characterized by their ability to mold easily. They can be found in everything from sound walls to telecommunications hardware to scuba goggles. Although they represent just under 2 percent of the global plastics market, they are notable in that they are made from a reaction between CO$_2$ and **epoxide** molecules. The presence of the catalyst, often zinc based, facilitates the **polymerization reaction** under very mild temperature and pressure, enabling CO$_2$ to be used as a feedstock for making polymers in a low-energy process. Polycarbonates have the capacity to store as much as half of their weight

Epoxide: a molecular compound with a characteristic ring structure comprising three atoms.

Polymerization reaction: a chemical reaction in which polymer "building blocks," or monomers, form polymer chains or networks.

---

* The terms *plastics* and *polymers* are not entirely interchangeable. Polymers are a general class of material comprising plastics, elastomers, and fibers, among other materials.

Polyol: an organic compound used in a wide variety of polymer chemistry.

Polyurethane: a polymer containing units joined together by carbamate compounds.

as CO$_2$. In addition, CO$_2$ and epoxides have been used for making polycarbonate **polyols** for **polyurethane**. The current global production of polyurethanes is around thirteen million tonnes with many applications in home furnishings and the automotive industry.

Emerging research is continuously demonstrating myriad possibilities for making polymers from CO$_2$. A recent study, for example, showed a new way of synthesizing block polymers, in which the CO$_2$ released in one polymerization reaction could be efficiently recycled into polycarbonate in a second reaction.[208] Indeed, the latest science is proving that incorporating CO$_2$ emissions into the production of polycarbonates to replace fossil-based polymers could provide a viable route toward reducing the carbon footprint of plastics production.

Many innovators and entrepreneurs around the world are taking on the challenge of decarbonizing the polymer industry. Mango Materials, based in California, is one of many companies that are rethinking the way in which plastics are manufactured and disposed of. Its technology produces a type of biopolymer, polyhydroxyalkanoate (PHA), using waste biogas via a microbial process. Normally waste sitting in landfills slowly decomposes, or is flared, to produce methane gas, which is then lost to the atmosphere. Al-

Anaerobic: characterized by the absence of oxygen.

Methanogenesis: the formation of methane by anaerobic bacteria.

ternatively, methane gas can be produced from waste, using organisms that perform **anaerobic** digestion, in a process known as **methanogenesis**. Mango Materials' process transforms methane sourced from anaerobically digested waste to make biodegradable PHB, which can go on to be used for making a wide variety of products, including clothing and packaging. The process forms a closed loop because the biodegradable products eventually decompose to make methane, and the cycle can continue.

Bio-inspired materials and processes are being increasingly adopted in polymer manufacturing and can offer a variety of economic and environmental benefits. One of the most important and growing applications of polymer materials is found in the automotive industry. Lightweight and sustainable materials can offer significant GHG emissions savings to transport vehicles by simply diminishing their weight and hence their gasoline requirements. Aluminum, magnesium, and carbon fiber-based materials have been at the forefront of lightweight materials for automotive applications; however, several new and exciting polymer-based materials are also emerging as promising candidates. These include composites, thermoplastics, thermoset polymers, rubbers, nanocomposites, and foams. In addition to the fuel-saving benefits drawn from their light weight, many of these emerging materials have the added appeal of being biologically based. Biofiber-reinforced composite materials, also known as biocomposites, for example, further reduce dependence on fossil sources by drawing instead from natural fiber compounds. Although there are still challenges to be overcome before biologically based materials can be found in all new vehicles, the future of biocomposites is certainly bright. The latest generation of biocomposite materials has achieved price and performance competitiveness with conventional automotive materials, and there is still potential to lower their manufacturing cost.[209] Their demonstrated economic and environmental viability has prompted interest from many of the world's largest automotive companies in investing in further biologically based material research.

Finally, the challenge of manufacturing high-quality plastics from plastic waste itself poses a barrier to the industry operating within a circular economy. Until now, plastic recycling has mostly been limited to mechanical sorting processes; however, recent advances in chemical recycling technology might help to recuperate plastic waste to create value-added products. Depolymerization can help to break down polymers into their raw constituents that can be used to make new polymers of similar quality. Mixed plastic waste can be transformed into naphtha, a liquid hydrocarbon used to manufacture a variety of chemicals, via pyrolysis. The impact of chemical recycling will ultimately be determined by the scale on which it is deployed, and policies will be required to incentivize plastic-waste recovery over the cheap production of plastic from fossil resources.

With innovative ideas and technologies emerging in the polymer industry every year, the story of plastics can hopefully become re-established as one that embraces both economic and environmental success.

Mindful Mineralization

When it comes to the carbon footprint of construction materials, cement production is the leading culprit. According to a recent report, the four billion tonnes of cement produced every year contribute to roughly 8 percent of all the world's CO$_2$ emissions.[210] The production of cement involves heating calcium carbonate (limestone) beyond 1,000°C in a large rotating kiln. The reaction decomposes the calcium carbonate into a solid calcium oxide product known as clinker, and CO$_2$ gas. The clinker is then collected, cooled, and ground for further processing. In addition to the emissions released in the reaction, the high temperatures needed to drive the process are derived from burning fossil fuels, thereby creating a second source of CO$_2$ emissions. This results in an average of 900 kg of CO$_2$ emitted to the atmosphere for every 1,000 kg of cement produced.

Concrete,* however, remains the most used construction material in the world, and abandoning or cutting back its production in the

---

* Although the terms *cement* and *concrete* often are used interchangeably, cement is actually one ingredient forming concrete, the others being aggregates (gravel or crushed stone) and water.

short term is unlikely. Clearly something must be done to address the industry's emissions problem. As it turns out, the process involved in making concrete offers a near picture-perfect opportunity to sequester CO$_2$ emissions.

The idea operates on the principle that gaseous CO$_2$ can be permanently mineralized in the form of carbonates. Making concrete begins with mixing water, aggregate, cement, and, often, additives. The resulting product is then transported to the work site or to a supplier of ready-mixed concrete. There the concrete is cured such that its moisture content is maintained to allow enough time for the water to react with the cement. This curing step is key to giving cement its strength and binding properties. Incorporating CO$_2$ into the concrete mixture can further enhance the concrete's strength by forming nanoscale calcium carbonate particles that act as nucleation sites for reaction of the water and cement and thus contribute to the material's strengthening. In the process, CO$_2$ is transformed into a highly stable carbonate mineral that is unlikely to revert back to its gaseous state.

Incorporating CO$_2$ into cement production makes sense from both an environmental and an economic perspective. Ready-mixed-concrete producers see an average strength improvement of approximately 10 percent when the concrete is injected with CO$_2$. Beyond fixing CO$_2$ as a stable mineral, incorporating CO$_2$ allows the concrete's cement content to be reduced by 5–8 percent without compromising its mechanical properties.

Many climate-conscious engineers are developing innovative strategies to decarbonize their cement and concrete processes. Formed in 2007, CarbonCure is a Canadian company that is actively incorporating recycled CO$_2$ emissions into its concrete processes. At full capacity its operation is anticipated to eliminate up to 500 Mt of CO$_2$ every year. Globally, it is estimated that CO$_2$ utilization for the concrete sector presents up to a USD 400 billion market opportunity, with the promise of reducing up to 1.4 Gt of CO$_2$ emissions annually by 2030.[211]

Ongoing research efforts are tasked to tackle the cement industry's massive carbon footprint. A 2019 paper reported a new approach to cement manufacturing that manages to mitigate CO_2 emissions and simultaneously produce value-added by-products.[212] The process uses an electrochemical cell, powered by renewable electricity, as well as pH gradients, to carry out the decarbonation of calcium carbonate and the precipitation of calcium hydroxide under controlled pH environments, thus generating CO_2, O_2, and H_2 gas. Reducing cement's carbon footprint using electrochemistry, however, will not be easy, considering the scale that would be required. Until the infrastructure exists, we may have to rely on decarbonizing current cement production.

As we have seen, capturing CO_2 in the form of a carbonate mineral lies at the heart of the cement industry's decarbonization strategy. It is worth pointing out, however, that mineralization also occurs naturally in the earth's carbon cycle. In this case, CO_2 from the atmosphere is captured by reacting it with metal oxides to form metal carbonates. The CO_2 mineralization of this kind can also be achieved in a processing plant or by injecting CO_2 into mineral-rich geological formations or aquifers. The chemistry involves reacting CO_2 with earth-abundant silicate minerals at relatively low temperatures. The formed metal carbonates are extremely stable, such that the reacted CO_2 cannot be easily released back into the atmosphere, thereby providing an excellent means of long-term sequestration.

Mineralization reactions are the exception to the notion that CO_2 chemistry cannot be performed without lots of energy, because they are actually energetically favored. In other words, CO_2 would rather exist in the form of a metal carbonate than as a CO_2 molecule. The fact that CO_2 is not located at the bottom of the thermodynamic valley is a good reminder that when it comes to transforming CO_2 emissions into product, we can work with nature and not against it.

Today carbon mineralization is not limited to the earth's natural carbon cycle. Carbfix, an Icelandic company, has developed an industrial process that takes captured CO$_2$ and stores it in the form of mineral carbon in subsurface rock. The technology allows 95 percent of the captured CO$_2$ to form rock within less than two years – which is a significant boost in mineralization rate compared to that occurring through the natural carbon cycle. In a similar spirit, a project led by the University of British Columbia is developing solutions to maximize the reaction of mine tailings that are rich in magnesium silicate – a waste by-product of mining operations – with captured CO$_2$ to form highly stable carbonate minerals. The project has already achieved mineralization on the time scale of weeks in the laboratory; however, the challenge remains of scaling the operation for use in the field. The solution is especially unique as it may not only serve as an effective tool for permanently storing carbon but also offer a solution for dealing with the waste tailings created by mining operations. According to Dr. Greg Dipple of the University of British Columbia, "reacting just 10 per cent of a mine's waste stream could be more than enough to offset the annual carbon emissions produced by a mining operation."[213] Although an industrial-scale CO$_2$-mineralization scheme for mine tailings has yet to be demonstrated, one can clearly appreciate the potential to render mining projects carbon neutral, using captured CO$_2$.

Sustainable Steelmaking

Steelmaking has been around since ancient times and has played a crucial role in the invention of key technologies, the development (and demise) of civilizations, and the path toward industrialization. Today steel production remains one of the most important industries in our global economy. Modern-day steel production, however, is worlds away from the first techniques developed by our ancestors. It took thousands of years to get from extracting ore to developing smelting techniques to optimizing steel's chemical composition. Even today we are still finding new and innovative ways to improve on modern steel-manufacturing processes.

At the heart of steelmaking is the reduction of iron ore to iron metal. From there it can be converted to myriad kinds of steel and subsequently cast into everything from slabs to sheets to rods to wires. What distinguishes iron metal from steel, however, is the presence of a **coke**. Interestingly, iron smeltering in Britain was first done with charcoal derived from wood; however, charcoal was quickly replaced with coked coal in the face of the country's growing wood scarcity.[214] The addition of coke, though small, makes a tremendous improvement to steel's mechanical properties, rendering it harder and stronger than pure iron metal. Steel can include other metal additives as well, depending on its application. Stainless steel, for example, owes its resistance to corrosion to the additional presence of chromium and nickel. Manganese, molybdenum, vanadium, silicon, and boron are other elements commonly found in steel alloys.

Coke: a carbon-rich compound that results from heating oil or coal in the absence of oxygen.

Transforming iron ore to iron metal requires a **reducing agent**. In a steel mill, iron ore is subjected to high temperatures in the presence of fossil-derived gases, such as synthesis gas, methane, or coal, whose carbon content enables it to reduce to iron metal. In addition to the emissions derived from the consumption of hydrocarbons in the steel mill, steelmaking carries a large carbon footprint from the energy needed to heat the furnaces, which typically operate at temperatures on the order of 1,500°C.

Reducing agent: a compound that can donate electrons to another chemical species.

Given the high-energy requirements of industrial ore production, the modern steelmaking industry is one of the largest industrial emitters. According to the International Energy Agency, the steel industry accounts for nearly 30 percent of direct carbon-dioxide emissions associated with the industrial sector worldwide. This is the equivalent to roughly 8 percent of the direct emissions from global fossil-fuel consumption.

Although the concept may appear promising, simply capturing the waste CO_2 emissions from steel mills is not a trivial solution. Capturing emissions would require costly installations, which pose a serious barrier to their adoption in existing plants. Furthermore, although renewable energy can be used, the amounts required could not be reliably supplied by the current renewable energy infrastructure. The emissions intensity of steel production cannot be readily resolved through incremental efficiency upgrades; as long as the reduction of iron ore is driven by fossil-sourced energy, and CO_2 remains a product of the reaction itself, steelmaking's notorious carbon footprint will not magically disappear.

There are, however, efforts to improve the situation by integrating the concept of carbon capture and utilization processes in steel mills using more economically viable means. The large quantities of available waste heat, coal, and steam produced in steel plants, for example, could be used to convert CO_2 into value-added products. A Germany-based project is already dedicated to exploring the use of the smelter gases from steel manufacturing in the creation of everything from fuels to plastics to fertilizer. They anticipate that their approach will address 10 percent of the annual carbon-dioxide emissions produced by Germany's industrial sector.[215] The Swedish steel mill Ovako, which already uses renewable energy to melt its scrap steel to produce its base product, has recently demonstrated that hydrogen can be burned instead of natural gas to produce heat for its rolling mills – without having any impact on steel quality.[216]

Co-opting industrial processes into steel manufacturing to take advantage of the high-energy process has also been proposed as an alternative approach to decarbonizing steelmaking. A recent techno-economic analysis,[217] comparing six different scenarios in which steel and methanol were coproduced, proposed that iron ore reduction via natural gas could offer a viable solution. The process involves reducing iron ore by using natural gas and by using the synthesis gas by-product to make methanol. The energy required by the reaction could theoretically be supplied from 100 percent renewable sources to render the process virtually CO_2 free. The proposed scenario could be adapted to conventional steel mills and so would not require new infrastructure. If applied in practice, the concept could mark a major step toward achieving carbon-neutral steel.

Although carbon capture and utilization strategies can certainly help to lessen the impact of steel production, the specifics of the best solution will not necessarily be identical for all steel mills, in all countries around the world. China, which is the largest steel producer, produces at a capacity that is nearly eight times that of the United States. China's steelmaking industry lies at the heart of the country's rapid economic growth.[218] The country alone is responsible for over 50 percent of the CO_2 emissions from global steel production, and its steel and iron manufacturing industry has rates of energy consumption and carbon emissions that are significantly higher than those of other countries.[219] As such, improving the processes associated with Chinese steel has the potential for major carbon-footprint reduction. A recent energy and economic analysis found that recovering waste heat from steel furnaces, to be used then for converting CO_2 to methanol, could potentially reduce the energy consumption and improve the profitability of Chinese steel mills.[219] Specifically it was estimated to yield a profit of USD 9.38 more per tonne compared to that from the conventional process. Alternative scenarios, such as employing a dry reforming process to make synthesis gas, despite yielding higher carbon-emission reductions, were not economically favorable due to the currently high cost of natural gas in China. From this example we can see that, when done properly, decarbonizing industrial processes, rather than posing economic risk, can improve efficiency and profitability. Region-specific studies such as these are therefore crucial in designing and improving our industries of the future.

No-Emission Nanotubes

Graphene is one of many **allotropes** of carbon, characterized by its two-dimensional honeycomb arrangement of carbon atoms, as shown in figure 18. It is the chemical and structural equivalent to a single sheet of carbon's more common allotrope, graphite. Graphene has been renowned in the world of materials for its incredible strength-to-thickness ratio and

Allotrope: a form in which an element can exist. For example, carbon can exist in many forms, including diamond, graphite, and charcoal.

its high thermal and electrical conductivity; it is anticipated to serve many applications, including flexible electronics, solar cells, and various chemical processes.

Carbon nanotubes, despite their fancy-sounding name, are simply graphene sheets that have been rolled up. They are most commonly made from the decomposition of fossil-derived organic compounds, such as methane or benzene, in the presence of a nanoparticle catalyst. As the organic compound decomposes, its constituent carbon atoms assemble on the catalyst, which serves as a template for the growth and formation of the carbon nanotube.

Carbon nanotubes currently have a commercial value of around USD 300,000 per tonne. Their unique set of mechanical and chemical properties render them ideal for a range of applications that include lithium-ion batteries, consumer electronics, and sporting goods. With so much technological potential, it would be ideal if carbon nanotubes could be manufactured from a fossil-free source. In a recent breakthrough[220] it was demonstrated that CO$_2$ from ambient air can be

Figure 18. **The honeycomb structure of graphene.**

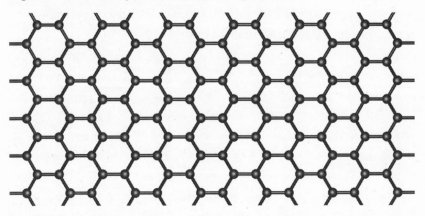

converted to carbon nanotubes via an electrochemical reaction. The process involves capturing CO$_2$ in the form of molten lithium carbonate, which can then be decomposed, or reduced, using an electric current to produce solid carbon and oxygen gas. The resulting carbon atoms assemble to form nanotubes with the help of a nickel catalyst. Different morphologies can be obtained by adjusting the relative amount of CO$_2$ and molten lithium in the cell. The carbon nanotube electrochemical cell can potentially be integrated into a natural gas electricity-generating power plant. The CO$_2$ emissions from the power plant could be fed to the carbon nanotube production unit, and the oxygen produced from the nanotube synthesis could be fed back into the power plant and used to improve the efficiency of the gas and steam turbines. It is estimated that the integrated natural-gas power plant and carbon nanotube–production facility could yield 8,350 kWh of electricity and 0.75 tonne of value-added carbon nanotubes with zero CO$_2$ emissions. By contrast, the power-generating plant alone yields 9,090 kWh of electricity and emits 2.74 tonnes of CO$_2$. The GHG-reduction potential derived by making carbon nanotubes from CO$_2$ augurs well for this technology's commercialization opportunities.

Mitigating Methane

With all this talk of CO$_2$ capture and conversion, it can be easy to forget climate change's second most important GHG culprit, methane. Like CO$_2$, methane emissions have increased dramatically since the industrial revolution, and, over a twenty year period, the global warming potential of methane is estimated to be about 85 times that of CO$_2$. The global warming potential of methane is estimated at about thirty times that of CO$_2$. At the current rate of increase the effect of anthropogenic methane on our climate will soon catch up and surpass today's unprecedented CO$_2$ concentrations. It is imperative that we begin to face the reality of our methane problem, in addition to our CO$_2$ problem.

Approximately 20 percent of the methane in our atmosphere stems from the production of fossil fuels, while another 50 percent comes from other anthropogenic sources, such as fermentation, cultivation, biomass burning, animal waste, sewage treatment, and landfills. The remainder emanates from natural sources that include permafrost, oceans, and termites. The 30 percent of methane that comes from natural sources may be especially difficult to control, particularly with the melting of permafrost that is anticipated to release huge quantities of trapped methane.

Every year around 140 billion cubic meters of natural gas are flared at thousands of oil fields globally. This results in more than three hundred million tonnes of CO$_2$ being emitted to the atmosphere, equivalent to emissions from approximately seventy-seven million cars. If this amount of gas were used for power generation, it could provide more electricity (750 billion kWh) than the annual consumption of the African continent. Currently, natural gas is flared for a variety of technical, regulatory, and economic reasons because capture is not given high priority. In the United States, technological advances in horizontal drilling and hydraulic fracturing led to a boom in shale gas extraction, allowing natural gas production to increase more than tenfold in just a decade.[221] Currently there exist around half a million natural gas wells and thousands of miles of pipelines, with no sign of the shale gas explosion slowing down. While there is some uncertainty in the exact amount of fugitive methane emitted from a natural gas well, it has been estimated that in CO$_2$ equivalents over a hundred-year timeline the effect of methane emissions in the United States will supersede the net total of all other GHG emissions from the country's iron, steel, aluminum, and cement manufacturing facilities combined.

Governments are clamping down on the practice of flaring, and yet oil producers are often left without economic options for dealing with this gas. The Zero Routine Flaring by 2030 initiative, which, at the time of writing, has been endorsed by thirty-two governments, thirty-seven oil companies, and fifteen development institutions, was launched in April 2015 by UN secretary-general Ban Ki Moon and World Bank president Jim Yong Kim.[222] The endorsers collectively represent more than 50 percent of global gas flaring. However, despite these efforts, the question still remains of how to deal with the rapidly increasing concentrations of methane building up in our atmosphere.

One strategy to tackle this challenge is to react simultaneously the two most potent anthropogenic GHGs (methane and CO$_2$) together in a process known as dry reforming. Like the previously described steam methane reforming, dry reforming produces synthesis gas; however, instead of using a mixture of methane and water, the process starts with a mixture of methane and CO$_2$. If both the methane and the CO$_2$ used were derived from the air or an industrial waste stream, dry reforming could work as a strategy of "killing two birds with one stone," while creating a useable chemical feedstock.

Although dry reforming has already become preferred as a means of mitigating GHG emissions, the process still requires lots of heat, which is often derived from the burning of fossil fuels. The ability to drive these high-temperature reactions using renewable energy, rather than fossil-fuel-derived means, can significantly reduce the energy-load requirements of the dry reforming process, rendering it more economical, energy efficient, and environmentally friendly in the long term.

It is also envisioned that direct sunlight could be used to drive the dry reforming reaction, thus providing an alternative approach to addressing the carbon emissions associated with heat production.[223] A solar approach to dry reforming also offers an advantage over the use of renewably generated heat, by offering better resilience to its infamous **coking** problem. In addition, dry reforming driven by sunlight could prove useful in remotely located natural gas fields.

> **Coking:** the formation of a carbon solid on the surface of a catalyst, which, in blocking its active sites, prevents the catalyst from functioning effectively.

Whether renewably generated electricity or direct sunlight is used, displacing the traditional heat sources that are often derived from the combustion of fossil fuels will be key to decarbonizing the dry reforming process. Furthermore, the capture and conversion of waste methane from landfills, anaerobic digesters, and waste treatment plants provides an opportunity to create value-added products and significantly reduce the long-term effects of climate change.

Carbon-Neutral Cooling

As the earth's temperature rises and local weather patterns change, many people may want to crank up their air-conditioning (AC). Finding ways of managing extreme temperatures is not new, and there is a long history of humans keeping themselves cool. In the prehistoric era, snow and ice provided cooling. Ancient Egyptians employed the evaporative cooling of moistened reeds hanging in open windows and exposed to a draft of air. Mechanical forms of evaporative cooling are traceable to second-century China.

The practice of winter ice harvesting and storage for the summer emerged in the seventeenth century. Willis Carrier of Buffalo, New York State, invented the first modern electrical air-conditioning unit in 1902. Considered one of the most useful inventions of the twentieth century, AC systems today pervade the modern lifestyle; they save lives, improve working conditions, make uninhabitable places habitable, and enable food preservation, computing operations, and many industrial processes.

The popularity of AC is growing rapidly, largely due to the growing global population that is enjoying income growth and increasing urbanization; however, as climate change forces the global temperatures to continue to rise, we can only expect AC use to increase. The number of AC units worldwide is expected to grow from 1.2 billion to 4.5 billion by 2050. The rising demand for AC will naturally be accompanied by an increase in electricity consumption and will have an impact on local climate as a result of the expelled heat generated in the compression-expansion refrigeration sequence. At this rate it has been estimated that the compounded effects of AC will be responsible for 0.5°C global warming by 2100.[224] Fortunately, however, there exists a real possibility that we will one day be able to switch on the AC without the guilt of being a shortsighted ancestor.

The vision of the AC system of the future is based on the observation that huge amounts of air pass through heating, ventilating, and AC (HVAC) systems. Even though the concentration of CO$_2$ in this air is low, the total amount of CO$_2$ that moves through these systems is substantial – possibly enough to warrant integrating CO$_2$ capture technology into HVAC systems. In considering ways in which we can remove excess emissions from air, it only seems natural to make use of our existing air-flow infrastructure.

A skeptic might point out that direct air-capture technology currently requires large amounts of energy, which, depending on its source, risks creating more emissions and, practically speaking, might not be available to residential and commercial buildings. One idea, however, is to perform on-site conversion of the captured CO$_2$ into liquid fuels and use the heat generated by the reaction to power the air-capture device. Of course, a source of H$_2$ generated from renewable energy via electrolysis is needed to ensure the process creates zero emissions.

While it may sound futuristic, we point out that all these technologies – direct air capture, water electrolysis, and CO$_2$-to-liquid-fuel conversion – exist and operate commercially. Moreover, integrating these processes into existing HVAC systems is claimed to be feasible, and a patent for a building-integrated CO$_2$-extraction system already exists. The on-site production of value-added fuels also offers huge economic benefits. According to one calculation, in a single hour twenty-five thousand grocery stores across Germany could potentially filter up to 40 kg of CO$_2$ out of the air and correspondingly produce 14 kg of liquid hydrocarbon.[225] And this is just from grocery stores; larger buildings are expected to have an even greater impact. On-site fuel generation could also revolutionize the way in which energy is distributed and managed. For example, owners of liquid-fuel-generating HVAC systems would effectively become **prosumers** and participate in an energy distribution grid system, analogous to that for renewable solar electricity. The need to decarbonize AC technologies could likely lay the infrastructural foundation for the next generation of advanced synthetic fuels.

Prosumer: a person who produces and consumes a product.

We started this section by talking about AC consumption and the need to keep cool. As it turns out, although CO$_2$ is contributing to the warming of our planet, it was one of the first refrigerants used in early cooling systems.

The first use of CO$_2$ as a refrigerant can be traced back to the mid-nineteenth century; however, CO$_2$, ammonia, and other first-generation refrigerants were later replaced by the more stable chlorofluorocarbons (CFCs) after the Second World War.

Under the 1989 Montreal Protocol, CFCs were phased out as a result of their effects on stratospheric ozone. They were replaced by perfluorocarbons (PFCs) and hydrofluorocarbons (HFCs), which, although they do not contribute to depletion of the ozone layer, are GHGs with high global warming potentials. The Kigali Amendment to the Montreal Protocol, which entered into force at the beginning of 2019, will phase out PFCs and HFCs and replace them with fourth-generation refrigerants, including ammonia, CO_2, propane, and tetrafluoropropene. These refrigerants do not deplete the ozone layer and possess relatively low global warming potential.

Carbon dioxide offers certain advantages over previous generation refrigerants, including higher volumetric cooling capacity, lower operating temperatures, non-flammability, and reduced operating costs. The only caveat is that CO_2 refrigeration systems need to operate at higher pressures, which results in them being more technologically demanding and costly compared to those systems based on HFCs. The high-pressure requirement, however, does come with certain advantages. The higher fluid density being circulated allows for cooling systems to be made more compact, which could open doors to the development of cooling systems for use in mobile applications, such as in cars and trucks. In addition, the high outlet temperature of the compressor could create heat utilization opportunities, including applications in automobile windshield de-icing and integrated home cooling and heating systems. In fact, the idea of using an automobile AC system with CO_2 as the refrigerant was proposed as early as 1989 by Professor Gustav Lorentzen. Nearly three decades later, and with vast research and development investments in these systems, Mercedes-Benz became the first to market a vehicle with a CO_2 air-conditioning system that offers improvements in fuel economy and in the speed of cooling and defrosting effectiveness and meets all pertinent automobile performance and safety standards. Volkswagen is expected to follow suit soon with an eco-friendly CO$_2$ air-conditioning system.

Carbon dioxide is returning as the refrigerant of the future as demonstrated by its burgeoning use in everything from ice- and snow-making machines and vending machines to cold storage rooms and supermarkets. Who said that CO$_2$ isn't cool?

KEY TAKEAWAYS

- Nearly all hydrocarbon products derived from fossil fuels can be obtained from methanol, which itself can be derived from CO$_2$ emissions and renewably sourced H$_2$.

- Using solar energy to power the Haber-Bosch process for making ammonia can help cut the carbon emissions associated with fertilizer production.

- Incorporating CO$_2$ emissions and bio-inspired materials and processes into the production of polymers can reduce the carbon footprint of plastics manufacturing.

- The cement and concrete industry's contribution to emissions can be addressed by injecting CO$_2$ into concrete, where it can be permanently mineralized in the form of carbonate.

- Steelmaking processes can be made greener and more efficient by using their vast amounts of waste energy to drive additional CO$_2$ utilization processes on site.

- Greenhouse gas methane and CO$_2$ emissions can be abated by reacting them together to make synthesis gas in a process powered by renewable energy or direct sunlight.

- Heating, ventilation, and air-conditioning systems could be integrated with a CO_2-capture device and a chemical reactor to perform on-site conversion of CO_2 into liquid fuels.

- Carbon dioxide is a viable and increasingly adopted refrigerant in wide-ranging cooling systems.

- Incorporating CO_2 emissions into the feedstock for industrial processes can offer both environmental and economic benefits.

- Although some CO_2-based technologies and processes are in the early stages of development, many of these concepts have already been commercialized (see appendix A for a list of companies currently practicing carbon utilization and related technologies).

Bringing It All Together

We hope that throughout these pages you have come to appreciate the potential of CO_2 to help our industries and technologies transition into the envisioned carbon-neutral economy of the future. Capturing CO_2, either directly from the air or from industrial waste streams, and sequestering it in highly stable forms, such as biochar or concrete, can help offset industry's stubborn emissions that cannot be resolved by renewable energy alone. In cases where a carbon capture and utilization strategy does not permanently sequester CO_2, the strategy can still help close the carbon loop by not introducing *new* emissions. For example, technologies that already use CO_2 as a feed, such as EOR and dry reforming, can make use of captured carbon emissions to fuel their processes, rather than drawing from fossil reserves. Similarly, captured CO_2, together with renewably sourced hydrogen, can replace fossil-derived feedstock to manufacture products such as methanol and polymers.

Beyond mitigating emissions, CO_2-based technologies can help support the growth of renewable energy and incentivize the development of a carbon capture infrastructure. For example, excess grid electricity can be used to convert CO_2 to fuels and chemicals, thus providing a storage option to help manage the intermittency challenge of renewable energy sources. Moreover, the capacity for CO_2 to be transformed into value-added products can render the economics of carbon capture technologies more favorable. While it is certainly not a silver bullet to climate change, integrating carbon capture, storage, and utilization with a renewable energy infrastructure can help meet, and preferably surpass, the emission targets set out by the Paris Agreement.

The exact degree to which CO_2 capture, storage, and utilization can contribute to limiting global temperature rise to 1.5°C depends largely on the associated supply chains, which makes it very difficult to predict at this point. For example, the carbon footprint of many of these processes depends largely on their access to an abundant supply of renewable energy. Furthermore, even if renewable energy were used to convert CO_2 into liquid fuels, such as methanol, subsequent combustion of the fuel would inevitably release carbon emissions into the atmosphere. Just because a process captures or uses carbon emissions does not guarantee that it is carbon neutral, let alone carbon negative.

Assessing whether or not a CO_2-based technology actually helps reduce emissions is critical to ensuring its potential to mitigate climate change. At the same time, for a solution to be successful, it must be economically viable and designed in the context of the local supply chains and social capital. For example, Iceland's infrastructure of methanol from CO_2 has a very low carbon footprint; however, this is the result of this country's abundance of renewably sourced electricity. The same scheme may not yield the same benefit in countries where the renewable energy infrastructure is less developed. Selecting the best strategy, therefore, becomes a question of which pathways have the best prospects in terms of mitigation, commercialization, scalability, environmental sustainability, social impact, and cost. The viability of any solution can only be assessed via a complete **life-cycle analysis**. This tool is often used by engineers and project managers to evaluate the energy, economic,

Life-cycle analysis: an assessment of the energy, economic, and environmental impacts of a product throughout its lifetime, cradle to grave – from the extraction of raw materials through to processing, transport, use, recycling, and disposal.

and environmental impact of a process or technology. The assessment is very thorough and typically starts at the extraction of source raw materials and carries through every stage in the processing, manufacture, transportation, distribution, use, maintenance, and ultimately disposal (and, hopefully, recycling) of the product. An analysis of this kind is crucial for identifying and evaluating the potential impacts associated with all energy and materials inputs and their outputs to the environment, making it more comprehensive in scope than a simple computation of the carbon footprint.

Throughout much of this discussion we will be equating concern for the environment with concern for climate change. In many ways this makes sense: climate change has major environmental repercussions by the nature of climate's ties to temperature and weather patterns, air composition, ocean levels, soil chemistry, biodiversity, and human activity. We should point out, however, that not all emission-mitigation strategies are environmentally friendly, and not all environmental conservation efforts help mitigate emissions. For example, on the one hand, biodegradable products are excellent from an environmental standpoint: a product that naturally degrades into the environment without releasing harmful substances can help reduce waste and minimize environmental impact. On the other hand, the degradation process may involve the release of carbon dioxide or methane into the atmosphere, which, if they are not captured to form a closed loop, is not ideal from the perspective of emissions mitigation. Alternatively, one could develop the world's best catalyst for converting captured CO_2 into polymers to make environmentally friendly plastics; however, if this catalyst were made from a toxic substance that risked being environmentally hazardous, it would probably have to be reconsidered. Performing a life-cycle analysis can help to evaluate these nuanced impacts of various "green" technologies.

Like any strategy striving for sustainability, reducing consumption must be central to CO_2 capture and utilization technology. This can be done by minimizing the amount of materials and energy needed for a process. For example, replacing a high-temperature industrial process with one that yields the same quantity of product at lower temperature can result in huge energy savings. In addition to consumption, the carbon footprint of the energy and materials feed also needs to be considered. Any heat or electricity used should ideally be derived from a renewable energy source. Another factor of particular importance to technologies that use CO_2 is the lifetime of the resulting product – that is, the length of time that the product holds carbon captive. This is extremely important when considering the impact of processes in which the primary strategy for negating emissions is sequestration.

In addition to offering environmental benefits, any viable technology or process should be scalable and economically efficient. These benefits are somewhat related in that the solution must have the potential to become widespread so that it may have as much impact as possible. Indeed, a new technology is useless if it is too complicated or too costly to be performed outside of a research laboratory.

Finally, the solution must be mindful of the safety and social well-being of everybody involved in its operation. Employing harmful materials or high-risk equipment, or failing to compensate workers with a living wage, would be in direct contradiction to the very goal of mitigation strategies, that is, the preservation of the environment and human life.

If at this point you are feeling overwhelmed by the various criteria used to evaluate the impact of CO_2-based technologies, remember that it is precisely this type of non-trivial framework that is needed for the design and implementation of truly sustainable solutions. This said, there are a few take-home messages to bear in mind. The first is that replacing fossil-derived feedstock with CO_2 is generally a good idea. The second is that the most direct incorporation of CO_2 into a product is most likely to yield the largest positive impact on its carbon footprint.[226] Finally, although generalized truths about carbon capture and utilization may be alluring, we must overcome the temptation of black-and-white thinking. Ultimately a full life-cycle assessment is needed to evaluate the environmental, economic, and social merits and the emissions-reduction potential of any solution.

Making It Happen

While many carbon capture and utilization technologies have proved to offer a safe and effective carbon-neutral solution, the achievement of an appreciable economic viability is often less obvious. With few CO_2-utilization technologies available on the market until recently, it remains more economical for many industries simply to emit the CO_2 and pay the carbon tax, rather than to capture and reuse it. Only with generous and consistent investment in research, accelerated market development, and radical policy reform can the vision of a global CO_2-utilization strategy be realized.

Increased government support for research can play a huge role in accelerating the development of key technologies associated with carbon capture, storage, and utilization. For example, improving the cost and efficiency of electrolysis devices to produce renewable H_2 will be key to rendering many CO_2-conversion processes viable. Similarly, a practical and economical electrochemical technology for reducing aqueous CO_2 would provide an attractive means to make a wide range of renewable chemicals and fuels. The same goes for improving the design and efficiency of carbon capture technologies. Chemistry and engineering, in particular, are the disciplines underpinning the discovery and optimization of catalytically active materials that facilitate the transformation of CO_2 to chemicals and fuels.

Consider a historical perspective on some energy-related, world-changing breakthroughs: the first practical solar cell is only a sixty-year-old story; the first practical light-emitting diode (LED) is fifty years old; and the conversion of light and/or electricity into energy-rich fuels has been seen, for the first time, during the past forty years. None of these problems have been simple, and all of these breakthroughs have relied on the unique physical and chemical properties of metals and semiconductors.

Despite our inevitable bias on this matter as scientists, we emphasize that science provides the foundation for many new and upcoming solutions. Making chemicals and fuels from CO$_2$ is a problem that requires science and engineering across multiple-length scales. Continued support for both basic and applied research is therefore vital to enable society to transition toward a net-zero emissions economy.

In addition to technological challenges, the success of carbon capture, storage, and utilization will require market development. Connecting existing processes with their appropriate supply chains, fostering collaborations between research institutes and start-ups, and creating support and partnerships for new companies all will help to accelerate the shared success of environmentally sustainable industries. Consumers also can play a major role in ensuring the creation of a vibrant market for CO$_2$-based technologies. Public endorsement, though crucial, is an often-underestimated element in the success of a new industry. Not only is public support necessary in the short term to foster investment into companies employing CO$_2$-utilization technologies, but also it is key to ensuring that government commits to a long-term carbon utilization policy.

Although personal opinions may differ as to the merit of different climate-change strategies (even among technical experts), unified support is needed to reduce dramatically our emissions as fast as possible, which requires, among other strategies, building widespread renewable energy infrastructure supplemented by carbon capture, storage, and utilization technologies.

Effective communication strategies are crucial for government and companies to gain public trust when it comes to carbon capture and utilization. On this note, we can learn from past instances in which a new technology, despite being promising and robust, gained widespread unpopularity among the public. A classic example is the story of genetically modified (GM) organisms. The media message in recent years has been clear: we must avoid GM foods at all cost if we value our health and the environment. While GM organisms undeniably pose a threat to biodiversity, they do however provide a means of ensuring our global food supply.[227] To further complicate matters, much of biotechnology became synonymous with multinational corporations that were notorious for exploiting small-scale farmers and forcing crop homogeneity. But the GM organisms themselves must not be confounded with the problematic history of their management, nor should their hypothetical risks be reason to cast aside their many benefits.[228] There exist few technologies that can be deemed "perfect" in terms of their environmental, climatic, economic, social, and cultural impact. We must embrace the nuances, weigh the options, and assess the risk based on our priorities at hand.

Carbon capture and storage have generally not been well received in the public eye. One study found that the negative view stemmed from the perceived risks associated with storage and transportation, despite their being touted as a promising strategy to achieving the 2°C warming target.[229] Interestingly, the findings revealed that carbon capture and utilization were viewed more positively, with perceived risks being related to the products themselves and their disposal. Overall, there was a low public awareness of carbon capture, storage, and utilization technologies, indicating that further communication of evidence-based knowledge on these technologies was needed. Fostering the healthy support from communities requires identifying the economic and social concerns of communities and ultimately fostering support of new technologies based on a history of trust.

Finally, no matter the existing state of technology or market space, support from government in the form of policy will be needed to help launch carbon-neutral and carbon-negative technologies. This is especially necessary in most countries and regions where the energy and industrial infrastructure is highly competitive and well established. Although one can be easily drawn into debates on the merits of free-market versus government-sanctioned approaches, a stable regulatory framework that is set in place by government is ultimately necessary in order for the private sector to thrive in a sustainable and equitable manner.

Advancing from technological readiness to actual commercial deployment can only be achieved if focused and informed policy measures are in place.[230] Policy in favor of climate-change mitigation can take various forms, although media tends to place disproportionate focus on carbon-pricing mechanisms, such as cap and trade and carbon taxes. Although putting a price on carbon emissions is necessary to encourage emission reductions, a complementary mix of ambitious policy containing significant regulatory elements will ultimately be required to meet short-term emission targets. Supportive policies that can enable the initiation, promotion, and growth of CO_2 products include government tax incentives and mandates, product labeling, government-supported certification and testing, and government oversight on life-cycle assessments. Policy must also reflect the longer timeline faced by clean-technology start-ups, whose scaling and infrastructure requirements can result in projects taking eight to ten years to launch.

The good news, however, is the consensus that CO_2 utilization can be commercially viable through sustained investment and dedicated policy measures. Bearing in mind that much of the existing energy infrastructure was previously, or is currently, heavily subsidized, subsidizing the emerging industries that will help guarantee a sustained transition from our current fossil-based economy seems to be a given.

Ultimately, the demonstration of technologically and economically sensible ways of fixing CO_2 to value-added chemicals and fuels, within a reasonable time frame, will breed confidence in further private and public investment; this investment will facilitate the turning of globally significant quantities of CO_2 from being a liability to being an asset. A recent study by the Global Carbon Capture and Storage Institute concluded that once the economic and technical feasibility of producing hydrocarbon fuels from CO_2 has been demonstrated, this could well accelerate the growth of carbon capture and sequestration and catalyze its mature commercial exploitation.[211] Knowledge dissemination through technological demonstrations will therefore be key to politicians and the public knowing about the global CO_2 utilization paradigm, which may ultimately help to inform policy decisions.

Envisioning the Economy of the Future

Our global community has been tasked to define and implement a whole-systems strategy for reducing CO_2 emissions at the giga-tonne scale. Accomplishing this goal requires a holistic paradigm that makes use of all the technologies in the renewable-energy and CO_2-utilization toolbox: CO_2-conversion reactors, water-electrolysis systems, water desalination, renewable-resource harvesting, and storage technologies. The integration of all these systems will be necessary for the sustained operation of an emissions-free economy, as shown in the flow diagram in figure 19.

What exactly will large-scale deployment of these systems look like? In truth, we do not have the answer, and the most appropriate solution is ultimately contingent on geographic location, available infrastructure, the amount of CO_2 at source, and the type and value of CO_2-based products compared to fossil-fuel-produced alternatives.

Capturing CO_2, either directly from the air or from industrial waste streams, can take many different forms. As mentioned earlier,

Figure 19. **The future energy economy** is one based on renewables and carbon capture, sequestration, and utilization.

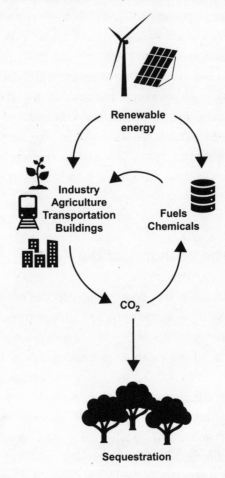

direct-air-capture units could be integrated into heating, ventilating, and AC systems in buildings, or as dedicated commercial operations, such as those of companies like Carbon Engineering. Where the captured CO$_2$ might be stored or used is highly dependent on its purity. For example, chemical manufacturing typically requires a high purity grade of CO$_2$ because the presence of impurities (i.e., any unwanted compounds that are filtered out along with the CO$_2$) can risk poisoning the catalyst; in the pharmaceutical industry a contaminated source risks rendering the product itself unusable.[231] So, although capturing high-purity CO_2 is more energy and cost intensive to obtain, it does pay off when it comes to utilization downstream. With this in mind, it is plausible to imagine that local need would have to be fulfilled by different grades of direct-air-capture devices – some are more expensive but able to produce pharmaceutical-grade product, and others are cheaper than, but as ubiquitous as, AC units.

How CO_2-conversion solutions might play out in practice is also subject to much imagination. The location of reactors might depend on proximity to renewable sources of H_2 and industrial-scale CO_2-capture facilities. Many questions remain as to the nature of the infrastructural changes that would be involved. Would the catalytic reactors be integrated within factories and refineries or located downstream from various industrial processes? And how would CO_2 be delivered to them? Would it be better for industrial operations to invest in on-site renewable power generation or to draw from the grid? In general, industries requiring large amounts of energy tend not to be reliant on the grid and instead employ their own power system. It may be that energy-intensive CO_2-utilization operations would install wind turbines and solar arrays dedicated to the processes, offsetting the capital expenditure costs against buying power.

The nature of the CO_2-conversion process – that is, the type of reactor needed – could also limit the location options. For example, electrochemically driven catalytic processes tend to be more compact compared to ones driven by heat. The versatility of electrochemical reactors could allow them to adopt modular designs, which could be assembled at different scales on demand. Micro catalytic reactors have the advantages of high-energy efficiency, conversion rates, and yields and provide scalable on-site demand and production with impressive process control.

Another vision is that of a modular CO_2 refinery in which various reactors convert CO_2, H_2, and methane into chemicals or liquid fuels. Hydrogen could be produced on-site from water by electrolysis, it could then be reacted with captured CO_2 or methane emissions to make synthesis gas, and the latter could ultimately produce fuels, such as methanol and gasoline. Already a spin-off company from Karlsruhe Institute of Technology in Germany, INERATEC, has commercialized innovative compact, containerized chemical plants, which enable these processes. They are modular CO_2 refineries that convert CO_2, H_2, methane, and water into both gaseous and liquid fuels. Their modular design, comprising units that are roughly the size of shipping containers, enables interlocking construction of CO_2 refineries at multiple-length scales that can be applied to various energy-related sectors.

A more futuristic vision includes a fleet of mobile CO_2 refineries that could be powered by renewable forms of electricity, built into truck-size containers, transported, delivered and integrated on-site to industries that have elected to convert their GHG emissions into synthetic fuels. Development of these portable, self-contained, modular synthesis, testing and control systems would enable renewable heat, electrical, or solar energy to be used at different industrial sites to capture and purify CO_2 and generate hydrogen from water electrolysis. It would indeed be revolutionary to imagine that instead of having to collect and transport industrial GHG emissions to a CO_2 refinery, one could deliver mobile CO_2 refineries to the emitting industries.

Rethinking our energy and industrial infrastructure also provides an opportunity to challenge conventional models of resource production and distribution. For example, the possibility of retrofitting air conditioners to convert water and CO_2 could redefine conventional approaches to energy ownership and distribution. Users could potentially collect the fuel for their own use and redistribute the excess in some form of local fuel-trading scheme. The concreteness and cooperative spirit of using renewably sourced energy to collect and convert CO_2 to value-added fuels in homes, apartments, and offices may be more likely to gain public support than more conventional large-scale industrial projects.[225] Indeed, local-level production would allow consumers to become more self-sufficient and empowered to participate in the carbon-neutral economy of the future.

Carbon capture and utilization is, of course, just one of many strategies to be undertaken to reduce our emissions, and it will be economically sustainable only if it is well integrated into the global energy landscape. Although not all countries have the technological capacity and resources to have their own refineries, all countries can still participate in the push toward synthetic fuels. For example, countries endowed with a robust renewable energy infrastructure can supply renewable electricity to countries that carry the CO_2-conversion infrastructure. There is some degree of the fossil-free solution appropriate for every location.

We began this story with the image of a chemical "tree" to conceptualize how all consumer products ultimately stem from a set of base resources. Today most chemicals necessary for the creation of everything from medicine to fertilizer to clothes still stem from fossil-derived resources. However, through these pages we hope that you have grasped the potential of substituting non-fossil alternatives for oil, natural gas, and coal. Returning to this idea, we leave you with a list of the key "root" chemicals and materials needed to sustain the manufacturing of critical commodities, and alternative procurement solutions to substitute their current fossil-fuel sourcing.

1. **Methanol** is a base chemical feedstock for nearly 30 percent of all chemical feedstocks, and more than 95 billion liters of methanol are produced each year. Today most methanol is still manufactured from syngas produced by catalytic reforming of fossil natural gas, and at least 0.6 kg of CO_2 is generated per

kilogram produced, depending on the specific process. As we saw earlier, however, methanol can be obtained directly, by catalytically reacting CO_2 with renewably sourced H_2 gas. Alternatively, it can be manufactured from syngas derived from biomass, specifically lignocellulose, to make bio-methanol.

2. **Ammonia** is key to securing the global food supply, and over 175 million tonnes is produced each year. It is a precursor to commonly used nitrogen-based fertilizers, such as urea, but can also be used directly. Ammonia is industrially produced via the energy-intensive Haber-Bosch process, in which hydrogen gas (typically derived from fossil sources) is reacted with nitrogen. This results in roughly 2.8 tonnes of CO_2 being emitted per tonne of ammonia produced – the process is responsible for 1 percent of all global CO_2 emissions. Reducing ammonia production's carbon footprint can be achieved by using a sustainable hydrogen source, adopting catalyst materials to lower the heat required by the reaction, and using renewable energy forms, such as sunlight, to drive the process. Ammonia can also be produced directly from water and nitrogen, effectively eliminating the need for hydrogen gas; however, this approach is still in its infancy and demands ongoing research efforts.

3. **Hydrogen gas** will be critical to a low-carbon future in many ways. It powers hydrogen fuel cells, is key to enabling much CO_2-conversion chemistry, and is also critical to many industries, particularly steel. As we saw earlier, sustainable hydrogen production can take several routes, including water electrolysis, biomass pyrolysis, or simply carbon capture integrated with steam methane reforming. In all cases, abundant and reliable sources of renewably generated electricity will be crucial to bring such operation to the scale required.

4. **Carbon monoxide** is critical to many major industrial processes, both on its own and as a key constituent of syngas. It is used widely in chemical manufacturing and metal refinement

and is vital to produce liquid hydrocarbons via the Fischer-Tropsch process. Carbon monoxide is produced industrially by several methods, the most common of which involves burning coke or other carbon-rich products that are derived from heating oil or coal. However, carbon monoxide can instead be obtained through the direct catalytic reduction of CO_2. A sustainable source of renewable energy, in the form of heat, electricity, or light, is needed to ensure that the process is carbon neutral, or even carbon negative.

5. **Olefins** form a key set of organic compounds, the most notable of which are ethylene and propylene. More than 150 million tonnes of ethylene is produced per year – more than any other organic compound. Olefins are the base ingredient of most plastics, surfactants, epoxides, and synthetic rubber. They are most commonly produced by heating natural gas or petroleum in a highly energy-intensive process known as steam cracking. Ethylene, however, can be obtained from bio-ethanol, which is produced from biomass, in lieu of petroleum products. Most recently, researchers have demonstrated that it can alternatively be obtained via the electrochemical catalytic reduction of CO_2.[232]

6. **BTX aromatics** are another critical set of organic compounds used in chemical manufacturing, the most important of which are benzene, toluene, and xylenes (hence, BTX). They are key to manufacturing plastics and polymers and are produced at a rate of over 100 million tonnes per year. The most common industrial route to obtain aromatics involves petroleum naphtha, a liquid hydrocarbon derived from crude oil. Like olefins, BTX aromatics can be produced from non-fossil sources, such as lignin from biomass, or by direct reduction of CO_2.[233]

7. **Carbon black** and **graphite** are both carbon-rich materials that are indispensable to the modern economy. Carbon black is used as a pigment and reinforcing filler in vehicle tires and

has many applications in both chemical manufacturing and electronics. Graphite, perhaps most famously associated with pencils, is used in many things, including steel, batteries, and nuclear reactors. Although technically different substances, carbon black and graphite are obtained by the incomplete combustion of fossil products, namely coal. Alternatively, carbon black can be made from biochar. Non-fossil alternatives to synthetic graphite production include mining; however, graphite recycling is the most sustainable route to eliminating the need for fossil resources. Researchers have recently demonstrated that carbonaceous species, similar to solid coal, can be obtained by catalytically reducing CO$_2$.[234] Scaling such a process could potentially eliminate our dependence on fossil resources for obtaining carbon-rich materials.

8. **Sulfur** is the principal precursor to sulfuric acid, a key substance for making fertilizer, pharmaceuticals, and various other chemicals. Over 180 million tonnes of sulfuric acid are produced every year; it is such an important chemical that the amount of sulfuric acid consumed by a country is considered a strong indicator of its industrial strength. While sulfur can be mined from natural sources, it is most commonly obtained as hydrogen sulfide from petroleum or natural gas. A non-fossil alternative, other than mining directly from the earth, would be to draw from biological waste sources of hydrogen sulfide, such as manure, wastewater, and biomass, coupled with the use of renewable energy.

9. **Silicon** is a critical material to modern industry. Though most commonly known for its use in semiconductor electronics, silicon in various forms is also of great importance to the metallurgical and chemicals industries. Silicon exists in abundance in the earth's crust, but the high-purity silicon required by industry is obtained by heating silica (the principal constituent of sand) in the presence of coke, a product of oil or coal. Finding a way to

produce industrial-grade silicon at lower temperatures without the need for a fossil-derived source of carbon could have significant impact on reducing the emissions associated with modern electronics. New research is proving that low-temperature silicon production is indeed possible. A recent study, for example, demonstrated a method to synthesize silicon nanowires in bulk quantities, using electrochemistry to reduce a calcium silicate mineral.[235] Alternatively, replacing fossil-derived coke with carbon-rich materials made from biomass or CO_2, and adopting renewable energy to drive the reduction of silica, could also eliminate the use of fossil resources in silicon production.

10. **Recycling**, though it does not parallel the rest of the items in this list, cannot go without a mention. Throughout this discussion we have emphasized the need to replace fossil-based processes with more sustainable, low-carbon alternatives that rely on renewable energy, biomass, and CO_2. However, we should point out that, wherever possible, repurposing, recycling, and reducing consumption of chemicals and materials – instead of producing them from scratch – offers major energy and emissions savings to our current manufacturing economy. In the analogy of our chemical tree, recycling translates to the reducing of the quantity of raw natural resources and root chemicals and materials that are needed to generate consumer goods. Key areas where recycling could have a real impact on emissions savings include the manufacturing of steel, plastics, paper, and lithium-ion batteries. Here, there are many opportunities for chemistry and engineering research to fill the gaps. For example, the creation of easily removable dyes can facilitate paper recycling, and alloy-separation technologies for scrap-metal recycling can eliminate the need to expand mining operations. Ultimately, reducing consumption – at both the individual and the industrial level – must lie at the heart of the low-carbon economy of the future.

The current chemical tree is not an innate structure. Like most human structures, it has been shaped by centuries of political, economic, technological, and social forces and therefore can and should be transformed to adapt to the needs of the present day. Indeed, it has already happened many times before. Between the two world wars the United States revised its entire chemical industry to protect its economy in anticipation of possible import contingencies. Ironically it was exactly during this time that it built up its chemical manufacturing around petroleum, natural gas, and agricultural by-products. With our current understanding of fundamental science and the expanse of engineering tools at our disposal, we cannot afford to limit our imaginations when it comes to what the chemical tree of the future might look like.

The Future Is Bright

The future of carbon-negative technologies is certainly bright. Economic indicators of the potential effects of climate change have drawn the attention of investors and financial regulators to the effect of carbon-emission controls on the stranded assets of fossil-fuel companies and to the impact of climate change on stock asset values. Over the past few years there has already been over a 10 percent increase in worldwide sales of clean technology. Estimates of global investments in innovative clean technology and enhanced resource efficiency amount to USD 2.9 trillion. This pales in comparison to the anticipated USD 90 trillion investments in low-carbon solutions for renewed urban, land, and energy infrastructure. It therefore comes as no surprise that "liquid fuels from sunshine," the aforementioned process in which sunlight-activated synthetic catalysts convert water and carbon dioxide into hydrocarbons, has been selected by the World Economic Forum's Expert Network and Global Future Councils in collaboration with *Scientific American* and its board of advisers as one of the top ten emerging technologies with the "potential to improve lives, transform industries, and safeguard the planet."[236]

In 2015, global aggregate investment in the renewable energy industry surpassed that in fossil fuels for the first time. It seems that the envisioned transition in our energy system, from non-renewable to renewable, is inevitable. Still, the challenge remains of meeting our emission targets quickly enough to avoid irreversible temperature rise. As put by the 2018 report of the Global Commission on the Economy and Climate, "the next 10–15 years are a unique 'use it or lose it' moment in economic history."[15] On this note, we should point out that adaption will become an increasingly critical part of a strategy going forward. This involves planning and preparing communities for the economic and environmental changes that lie ahead. Even if global temperature rise were to be successfully limited to 1.5°C, we are already seeing that it is necessary to adapt to the effects of the mere 1°C increase we are currently experiencing. Extreme-weather preparedness will be particularly critical to the preservation of human life in both the short and the long term.

We are nevertheless beginning to see encouraging signs of change on the horizon. The global energy revolution has begun, with clean-technology solutions increasingly contributing to our daily lives. These solutions are exemplified by energy-efficient lighting, electric cars, solar and wind electricity generation, large-scale battery storage, hydrogen fuel cell buses, smart energy-saving windows, solar thermal heating systems, green cement, and self-cleaning buildings. Furthermore, CO_2-utilization technologies are becoming more than just a musing of academic research papers. Appendix A provides a long, though not complete, list of companies around the world that are already making fuels, chemicals, and consumer products out of CO_2. We encourage you to learn and support these creative and important endeavors; indeed, they are leading our energy revolution and paving the way toward a zero-emission economy.

We have also compiled a list of ten actions you can take to lessen the impact of climate change:

1. **Educate** those around you about the need for action on climate and the vision of an industrial and energy infrastructure that operates with net-zero emissions. One does not need to be an authority on climate science or policy to communicate the urgent need for climate action; simply talking about it sends a message of seriousness that is effective in itself. Moreover, we must not underestimate our individual influence within our networks and communities. The success of renewable energy technologies and recycling initiatives are largely thanks to widespread public knowledge of them. Help spread the message!

2. **Vote** in favor of climate-change mitigation. This is arguably the most important step that a citizen can take. Government commitment to achieving national and regional emission targets, as well as leading and participating in the international climate-change conversation, is critical to ensuring effective mitigation and adaptation strategies.

3. **Write** to your elected representatives, urging them to support and commit to emission-reduction strategies in your community. These can include investing in public transportation infrastructure, implementing pollution pricing schemes, and creating policy to support and sustain an energy landscape based on 100 percent renewable sources and carbon-negative technologies.

4. **Read** more about the carbon footprint associated with various products and activities. Did you know, for example, that bananas are one of the most carbon-friendly foods?[55] Keeping informed of the impact of our daily lives on our planet is key to the individual decision-making that can eventually lead to widespread changes in consumer attitudes. A list of recommended reading material is provided in appendix B.

5. **Invest** in carbon-neutral and carbon-negative technologies (if it is financially feasible for you, of course). Many companies based on renewable-energy and carbon-negative technologies have appeared on the scene in recent years, and supporting their entry to market is key to ensuring their success. A list of existing companies is provided in appendix A. If you want the diversity of broad indexes such as the S&P 500, fossil-free variants are available as exchange-traded funds.

6. **Eat** less red meat.* You do not need to become vegan to have an impact on the planet. Reducing meat consumption alone is one of the highest-impact actions that individuals can take to reduce their carbon footprint.[237] The recovery of pasture lands for natural vegetation would significantly increase carbon uptake from the atmosphere, as well as reduce the methane and nitrous oxide emissions that result directly from cattle. It goes without saying that shifting toward a plant-based diet could have far-reaching benefits to the environment and the climate. Also important is the need to reduce our food waste. It is estimated that one-fifth of all the food produced in Canada annually ends up in a landfill, or alongside organic waste – that is, food that could have otherwise been eaten.[238] Although the waste occurs at various points along the food supply chain, individual households can have a real impact on reducing food waste. If you are not convinced, consider that being mindful about planning your grocery needs, freezing food, and saving leftovers is also good for your wallet.

7. **Travel** consciously. Walking, cycling, and public transit are ideal choices for commuters when it comes to their carbon

---

* This is a general guideline that should only be considered by individuals whose financial and environmental circumstances guarantee their access to nutritious food. Furthermore, an individual's dietary choices should not be subject to any rules or restrictions that pose a risk to their physical and/or mental well-being.

footprint. As these options are not necessarily available or viable for those living in rural areas, this is an even better reason their lobby government to invest in public transportation infrastructure!

If a motor vehicle is necessary to your lifestyle, switching to a more energy-efficient engine and choosing to carpool are effective strategies for reducing your carbon footprint. Purchasing a hybrid or electric vehicle is another excellent option if battery-charging infrastructure is available in your region. As renewable energy infrastructure is further developed, and charging stations that provide 100 percent renewably sourced electricity are established, we can expect that electric vehicles will become an increasingly important technology in reducing emissions. Finally, "flight shaming" (and any kind of shaming, for that matter) will not solve our climate crisis. We must recognize that it will take time for procurement structures to change.

8. **Assess** energy consumption in your home. According to one environmental assessment study, commodities and energy use associated with households are connected to supply chains that are responsible for up to 60 percent of GHG emissions worldwide.[239] Although a household's transportation habits were the greatest determinant of their carbon footprint (see point 7), consumption associated with shelter, such as the burning of household fuel, was the second-largest source of direct emissions. Reductions in GHG emissions from building retrofits can be achieved faster and tend to be more cost-effective compared to other climate-mitigation measures. Dr. Christina Hoicka of York University and Dr. Runa Das of Royal Roads University argue that retrofitting buildings with the goal of maximizing social and environmental benefits is an underestimated strategy that can help meet emission targets in the short term.[240] So, be mindful of your lighting and thermostat habits and check to

ensure that all your home appliances are running efficiently. If possible, seek a third-party energy assessment of your household to help identify ways in which you can improve the energy efficiency of your home.

9. **Reduce** consumption wherever possible. Changing consumer behavior is a critical factor in ensuring sustainable management of our resources in the long term. This is becoming increasingly important as the fastest-growing populations strive for a Western standard of living that is characteristically carbon intensive. Remember that your use of a product comprises just a small fraction of the carbon emissions associated with its full life span. An innocent-seeming disposable coffee cup, though not spewing carbon dioxide into the atmosphere as you hold it, has emissions associated with the extraction of raw materials, the manufacturing of polyethylene plastic used to line its interior, and the vehicle that brought it to the café, not to mention the fumes it will emit from the landfill. So, be mindful of the everyday products that you purchase and consume, and remember that individual action begets collective action.

Although little research has been carried out on the relative carbon footprint of online versus traditional shopping, some recent studies are beginning to shed light on the matter. The answer is not clear, however, and depends greatly on the method of transportation available to the consumer, as well as the frequency of the purchases.[241] Although online shopping generally has a smaller carbon footprint compared to regular shopping, next-day or two-day shipping (based on a US study) may not be the same, because it adds stress to logistical systems, often forcing deliveries to be made via non-optimal transport pathways.[242] The situation in other countries could be different, particularly when distances are smaller and aircraft are not used for the same-day or two-day deliveries. So, next time you are tempted by the promise of two-day shipping at no extra charge, remember that we all have a role to play in protecting our planet, and opt for the regular shipping instead.

10. **Support** evidence-based decision-making. Many existing networks are doing great work in lobbying government and policymakers to turn to science to help create effective climate policy. For more information, visit

- Campaign for Science and Engineering (United Kingdom), http://www.sciencecampaign.org.uk/
- Citizens' Climate Lobby (Canada and United States), https://citizensclimatelobby.org/
- Evidence for Democracy (Canada), https://evidencefor-democracy.ca/
- Indigenous Environmental Network (United States and Canada), http://www.ienearth.org/
- Sunrise Movement (United States), https://www.sunrisemovement.org/
- 350.org (International), https://350.org/
- Toronto Science Policy Network (Canada), https://toscipolicynet.ca/
- Union of Concerned Scientists (United States), https://www.ucsusa.org/

Since it was first created from carbon and oxygen atoms after the big bang, carbon dioxide has been on an interesting journey, as a constituent of an atmosphere unable to support life, to its substitution by oxygen molecules that are able to sustain life. Working together in a carbon cycle, CO$_2$ and oxygen have allowed all life on earth to flourish, but we are beginning to realize that too much of one and too little of the other can bring a troubling end to this delicate balancing act.

We hope that throughout these pages you have become convinced that incorporating captured CO$_2$ into our industrial processes and technologies can provide alternative carbon-neutral solutions to our existing fossil-fuel conundrum. Change is happening, and it is hoped that the pace of the energy transition will be fast enough to keep below the tipping point of anthropogenic climate change that is stimulated by carbon dioxide. Ultimately deep systemic changes in our economic structures will be needed to ensure a sustainable and equitable future for all, and social movement coalitions must be at the forefront of these transformations.

The exciting potential that carbon dioxide capture, sequestration, and utilization technologies offer to help avoid this scenario, with gigantic rewards to the economy, environment, and climate, is our story's central message. It is a tall order for society, but we are all in this world together, and we must all now shoulder the Herculean responsibilities of caring for the earth and our collective future with the same degree of concern that we have for ourselves.

KEY TAKEAWAYS

- A process that captures and/or uses CO_2 emissions is not necessarily carbon negative or carbon neutral. The carbon footprint associated with all its energy and materials inputs and outputs must be considered to determine its net emissions.

- A life-cycle analysis is a cradle-to-grave evaluation of the energy, economic, and environmental impact of a technology or process.

- To be successful, new technologies should be scalable, economically viable, socially responsible, and environmentally benign.

- The vision of a global CO_2-utilization strategy can be realized with generous investment in research, accelerated market development, and radical policy reform.

- There is consensus that CO_2 utilization can be commercially viable with short turnaround times for investment returns.

- Effective and transparent communication is crucial to gaining public trust and support for CO_2 capture, storage, and utilization technologies.

- Knowledge dissemination through technological demonstrations and pilot operations can help educate non-experts on the emission-free alternatives to fossil-based processes.

- Large-scale deployment of CO_2 capture, storage, and utilization technologies is contingent on geographic location, available infrastructure, accessibility to renewable energy, local market, and competition posed by fossil-based technologies.

- Rethinking our energy and industrial infrastructure provides an opportunity to challenge conventional models of resource production and distribution.

- Adaption, in the form of planning and preparing communities for economic and environmental changes, will become an increasingly critical part of climate action strategies going forward.

Appendix A
Companies Transforming CO_2

AGG Biofuel
United States
Converting CO_2 and other carbonaceous inputs into syngas for use in transportation fuels, power generation, process heat, and added-value products.

The Air Company
Canada and United States
Converting CO_2 to ethanol to make carbon-neutral alcoholic beverage products. https://aircompany.com/

Algenol Biofuels
United States
Converting algae biomass into a variety of sustainable products, such as fertilizer, colorants, and nutritional supplements. http://algenol.com/

BioBTX
Netherlands
Converting sustainable feedstock and plastic waste to BTX aromatics. https://biobtx.com/

Blue Planet
United States
Capturing and converting CO_2 into aggregates for building and highway materials via a carbon-mineralization process. http://www.blueplanet-ltd.com/

C2CNT

Canada and United States

Capturing CO_2 from the atmosphere or from industrial flue stacks and converting it into carbon nanotubes via electrolysis. http://www.c2cnt.com/

C4X

China, Canada, and United States

Capturing CO_2 from industrial flue gas sources to make ethylene carbonate, ethylene glycol, methanol, and low-density foams. http://www.ccccx.net/en/index.asp

Carbicrete

Canada

Making high-quality, carbon-negative concrete out of steel slag (a by-product of the steelmaking process) and CO_2. http://carbicrete.com/

Carbon Clean Solutions

United Kingdom, India, and United States

Providing low-cost CO_2-separation technology for industrial and gas treating applications. https://carboncleansolutions.com/

CarbonCure

Canada

Injecting industrial waste CO_2 into ready-mixed concrete and concrete masonry to enhance their quality, while sequestering carbon. https://www.carboncure.com/

Carbon Engineering

Canada

Capturing CO_2 from air to deliver a purified compressed stream of CO_2, as well as converting CO_2 into synthetic transportation fuels. http://carbonengineering.com/

Carbonfree Chemicals
United States

Capturing CO_2 from industrial flue gases and converting it into solid carbonate materials to create products such as sodium bicarbonate (baking soda), hydrochloric acid, caustic soda, and household bleach. http://www.carbonfreechem.com/

Carbon Recycling International
Iceland

Capturing CO_2 released by a geothermal power plant and converting it to methanol, using renewably generated electricity and hydrogen. http://www.carbonrecycling.is/

Carbon Upcycling Technologies
Canada

Converting CO_2 with waste products, such as fly ash, graphite, coal, or pet coke, to create nanoparticles that are used to enhance the performance of concrete, plastic, and coatings. http://carbonupcycling.com/

Catalytic Innovations
United States

Converting CO_2 into high-purity alcohols via electrochemistry. http://www.catalyticinnovations.com

Climeworks
Switzerland

Removing and concentrating CO_2 from thin air via a cyclic heating filtration system. http://www.climeworks.com/

Connovate
Denmark and Philippines

Making long-lived building construction panels, using less cement. https://www.connovate.com/

Cool Planet
United States

Sequestering CO_2 to make biochar-based products that are used to improve soil quality and enhance plant growth. https://www.coolplanet.com/

CO_2 Solutions
Canada

Capturing and purifying CO_2 from industrial sources for reuse, conversion, or storage. https://co2solutions.com/

Covestro
Germany

Manufacturing polyols out of CO_2 to create foam products. https://www.co2-dreams.covestro.com/

Dimensional Energy
United States

Converting CO_2 into polymers and chemical intermediaries via photochemistry. https://www.dimensionalenergy.net/

DyeCoo
Netherlands

Using recycled CO_2, instead of water, as the dyeing medium to provide clean textile-processing solutions on an industrial scale. http://www.dyecoo.com/

Ecoera
Sweden

Sequestering carbon-rich agricultural residues to produce biochar used for soil enhancement. https://ecoera.se/

Econic Technologies
United Kingdom

Producing polyols by reacting CO_2 with epoxides to manufacture a range of sustainable polymer products. http://econic-technologies.com/

Electrochea
Germany and Denmark

Converting CO_2 to natural gas via biocatalysis, using electricity. http://www.electrochaea.com/

Enerkem
Canada

Converting non-recyclable waste to biofuels and chemicals. https://enerkem.com/

Global Thermostat
United States

Capturing and purifying CO_2 directly from the atmosphere, as well as from industrial flue streams, using only steam and electricity. https://globalthermostat.com/

Green Minerals
Netherlands

Capturing and converting CO_2 into concrete, paper, and polymers via a mineralization process. http://www.green-minerals.nl/

INERATEC
Germany

Converting CO_2 into syngas, methane, and liquid fuels, such as methanol. https://ineratec.de/

Innovator Energy
United States

Capturing CO_2 from flue gas, using power-plant condenser water. https://innovatorenergy.com/

Kiverdi
United States

Converting CO_2 into high-value oils, nutrients, and bio-based products, using a biocatalytic process involving hydrogenotrophs, organisms that metabolize hydrogen. https://www.kiverdi.com/

LanzaTech

Global

Converting CO_2 into ethanol, jet fuel, and a variety of chemicals, using a biocatalytic process based on gas-fermenting organisms. http://www.lanzatech.com/

Liquid Light

United States

Converting CO_2 into a variety of industrial chemicals, using an electrochemical process. http://llchemical.com/

Mango Materials

United States

Converting methane gas emissions from waste facilities into biopolymers, using a microbial process. http://mangomaterials.com/

Net Power

United States

Converting fossil fuels into electricity in a zero-emission process at a lower cost than that of existing power plants. https://www.netpower.com/

Newlight Technologies

United States

Capturing and converting CO_2 from industrial sources to make bioplastics, using a biocatalytic process. https://www.newlight.com/

Opus12

United States

Converting CO_2 to fuels and chemicals, using electrocatalysis. https://www.opus-12.com/

Photanol
Netherlands

> Converting CO$_2$ to organic acids for the production of a range of biochemical products, using genetically modified cyanobacteria and sunlight. https://www.photanol.com/

PhotoSynthetica
United Kingdom

> Developing building-cladding systems that convert CO$_2$ to biomass, using algae. https://www.photosynthetica.co.uk/

Plasco Conversion Technologies
Canada

> Using recycled heat to gasify industrial waste, which is then converted to syngas. http://plascotechnologies.com/

Pond Technologies
Canada

> Converting CO$_2$ to valuable algae-based products using bioreactors coupled with LED lighting systems. https://www.pondtech.com/

SkyMining
Sweden

> Capturing CO$_2$ from the atmosphere in the form of biomass, which is then converted to biofuel. https://www.skymining.com/

Skytree
Netherlands

> Capturing and converting CO$_2$ directly from the atmosphere into methanol, using biocatalytic processes on both small and large scale. https://www.skytree.eu/

Solidia Technologies

United States

Sequestering CO_2 by using it as an additive to enhance the performance of concrete. http://solidiatech.com/

Solistra

Canada

Converting CO_2 to chemicals and fuels by gas-phase heterogeneous photocatalysis. http://www.solistra.ca

Sunfire

Germany

Converting CO_2 and renewably generated H_2 into syngas, and subsequently into synthetic natural gas, liquid fuel, and other chemical products, using electricity. https://www.sunfire.de/en/

Syzygy Plasmonics

United States

Developing photocatalysts and photocatalytic reactors to enable different high-value products, with a focus on hydrogen production. http://plasmonics.tech/

TerraCOH

United States

Using CO_2 as a geothermal working fluid in lieu of water for lower-cost, higher-efficiency geothermal electricity production. http://www.terracoh-age.com

Appendix B
Further Reading

Aronoff, K., Battistoni, A., Cohen, D. A., & Riofrancos, T. *A Planet to Win. Why We Need a Green New Deal* (Verso, 2019).

Berners-Lee, M. *How Bad Are Bananas? The Carbon Footprint of Everything* (Green Profile, 2010).

Buck, H. J. *After Geoengineering: Climate Tragedy, Repair, and Restoration* (Verso, 2019).

Favaro, B. *The Carbon Code. How You Can Become a Climate Change Hero* (Johns Hopkins University Press, 2017).

Harvey, H., Orvis, R., & Rissman, J. *Designing Climate Solutions. A Policy Guide for Low-Carbon Energy* (Island Press, 2018).

Kalmus, P. *Being the Change. Live Well and Spark a Climate Revolution* (New Society Publishers, 2017).

Kawken, P. *Drawdown. The Most Comprehensive Plan Ever Proposed to Reverse Global Warming* (Penguin Books, 2017).

Klein, N. *This Changes Everything* (Simon & Schuster, 2014).

Kolbert, E. *The Sixth Extinction* (Henry Holt and Co., 2014).

Maxton, G. & Randers, J. *Reinventing Prosperity. Managing Economic Growth to Reduce Unemployment, Inequality and Climate Change* (Greystone Books, 2016).

Stokes, L. C. *Short Circuiting Policy. Interest Groups and the Battle Over Clean Energy and Climate Policy in the American States* (Oxford University Press, 2020).

Vollmann, W. *Carbon Ideologies* (Viking, 2018).

References

1 Macvean, M. For many people, gathering possessions is just the stuff of life. *Los Angeles Times* https://www.latimes.com/health/la-xpm-2014 -mar-21-la-he-keeping-stuff-20140322-story.html (21 March 2014).

2 Keeling, R. F. *et al.* Atmospheric Monthly in situ CO_2 Data – Mauna Loa Observatory, Hawaii. In Scripps CO_2 Program Data. UC San Diego Library Digital Collections. https://doi.org/10.6075/J08W3BHW (accessed, 7 February 2020).

3 Meure, C. M. *et al.* Law Dome CO_2, CH_4 and N_2O ice core records extended to 2000 years BP. *Geophysical Research Letters* **33** (2006).

4 Griffin, D. P. CDP carbon majors report 2017. https://www.cdp.net/en /reports/archive (accessed, 21 January 2019).

5 World Meteorological Organization. WMO statement on the state of the global climate in 2019. https://public.wmo.int/en/resources/library /wmo-statement-state-of-global-climate-2016 (accessed, 2020).

6 World Meteorological Organization. WMO statement on the state of the global climate in 2018. https://library.wmo.int/doc_num.php?explnum _id=5789 (2019).

7 Krishnan, R. *et al.* Unravelling climate change in the Hindu Kush Himalaya: rapid warming in the mountains and increasing extremes. In *The Hindu Kush Himalaya Assessment. Mountains, Climate Change, Sustainability and People* (eds Wester, P., Mishra, A., Mukherji, A. & Shrestha, A. B.) 57–97 (Springer International Publishing, 2019). doi:10.1007/978-3-319-92288-1_3.

8 World Meteorological Organization. 2017 is set to be in the top three hot-test years, with record-breaking extreme weather. https://public.wmo .int/en/media/press-release/2017-set-be-top-three-hottest-years-record -breaking-extreme-weather (2017).

9 World Meteorological Organization. WMO climate statement: past 4 years warmest on record. https://public.wmo.int/en/media

/press-release/wmo-climate-statement-past-4-years-warmest-record (2018).

10 Internal Displacement Monitoring Centre. Global report on internal displacement, 2018. https://www.internal-displacement.org/sites/default /files/publications/documents/201805-final-GRID-2018_0.pdf (2018).

11 IEA. Global energy and CO_2 status report, 2019. https://www.iea.org /reports/global-energy-co2-status-report-2019 (IEA, 2019).

12 CAIT Climate Data Explorer via Climate Watch. Total greenhouse gas emissions including land use change and forestry, measured in tonnes of carbon dioxide-equivalents 1990–2016. https://www.climatewatchdata. org/data-explorer/historical-emissions (accessed, 15 November 2019).

13 The emissions gap report 2017: a UN environment synthesis report. https://www.unenvironment.org/resources/emissions-gap-report-2017 (2017).

14 Rogelj, J. et al. Understanding the origin of Paris Agreement emission uncertainties. Nature Communications 8, 15748 (2017).

15 The Global Commission on the Economy and Climate. The new climate economy. https://newclimateeconomy.report/2018/ (2018).

16 Victor, D. G. et al. Prove Paris was more than paper promises. Nature News 548, 25 (2017).

17 Pain, S. Power through the ages. Nature 551, S134 (2017).

18 IEA. Global energy and CO_2 status report, 2019: Analysis. https://www .iea.org/reports/global-energy-and-co2-status-report-2019/emissions (IEA, 2019).

19 Howarth, R. W. Methane emissions and climatic warming risk from hydraulic fracturing and shale gas development: implications for policy. Energy and Emission Control Technologies https://www.dovepress.com /methane-emissions-and-climatic-warming-risk-from-hydraulic -fracturing--peer-reviewed-article-EECT (2015) doi:10.2147/EECT.S61539.

20 Friedlingstein, P. et al. Global carbon budget 2019. Earth System Science Data 11, 1783–1838 (2019).

21 Psarras, P. et al. Slicing the pie: how big could carbon dioxide removal be? Wiley Interdisciplinary Reviews: Energy and Environment 6, e253 (2017).

22 Gerber, P. J. et al. Tackling climate change through livestock: a global assessment of emissions and mitigation opportunities. Food and Agriculture Organization of the United Nations (FAO), Rome (2013).

23 Griscom, B. W. et al. Natural climate solutions. PNAS 114, 11645–11650 (2017).

24 Bastin, J.-F. et al. The global tree restoration potential. Science 365, 76–79 (2019).

25 Veldman, J. W. *et al.* Comment on "The global tree restoration potential."
    Science **366**, eaay7976 (2019).

26 Meyfroidt, P. *et al.* Forest transitions, trade, and the global displacement
    of land use. *Proceedings of the National Academy of Sciences* **107**, 20917–
    20922 (2010).

27 Holl, K. D. & Brancalion, P. H. S. Tree planting is not a simple solution.
    Science **368**, 580–581 (2020).

28 Thompson, A. The new plants that could save us from climate change. *Pop-
    ular Mechanics* https://www.popularmechanics.com/science/green-tech/
    a14000753/the-plants-that-could-save-us-from-climate-change/ (2017).

29 de Jong, E. & Jungmeier, G. Biorefinery concepts in comparison to petro-
    chemical refineries. In *Industrial Biorefineries and White Biotechnology*, 3–33
    (Elsevier, 2016).

30 Friedmann, S. Julio, Fan, Z. & Tang, K. *Low-Carbon Heat Solutions for
    Heavy Industry. Sources, Options, and Costs Today.* New York (Center on
    Global Energy Policy, 2019).

31 Dahlmann, K. *et al.* Climate-compatible air transport system: climate
    impact mitigation potential for actual and future aircraft. *Aerospace* **3**,
    38 (2016).

32 Scherer, L. & Pfister, S. Hydropower's biogenic carbon footprint. *PLOS
    ONE* **11**, e0161947 (2016).

33 Kennedy, M., Mrofka, D. & von der Borch, C. Snowball Earth termination
    by destabilization of equatorial permafrost methane clathrate. *Nature* **453**,
    642–645 (2008).

34 Sengupta, A. & Gupta, N. K. MWCNTs based sorbents for nuclear waste
    management: a review. *Journal of Environmental Chemical Engineering* **5**,
    5099–5114 (2017).

35 Corless, V. How do we power a sustainable future? *Advanced Science News*
    https://www.advancedsciencenews.com/how-do-we-power-a-sustainable
    -future/ (2019).

36 Thurner, P. W., Mittermeier, L. & Küchenhoff, H. How long does it take
    to build a nuclear power plant? A non-parametric event history approach
    with P-splines. *Energy Policy* **70**, 163–171 (2014).

37 Cho, A. The little reactors that could. *Science* **363**, 806–809 (2019).

38 Mental-health effects of the Chernobyl disaster live on. *The Lancet* **366**,
    958 (2005).

39 World Health Organization. Chernobyl: the true scale of the accident.
    http://www.who.int/mediacentre/news/releases/2005/pr38/en/ (2005).

40 Domingo, T. *et al.* Fukushima-derived radioactivity measurements in
    Pacific salmon and soil samples collected in British Columbia, Canada.
    Can. J. Chem. **96**, 124–131 (2017).

41 Schrope, M. Nuclear power prevents more deaths than it causes. *Chemical & Engineering News* https://cen.acs.org/articles/91/web/2013/04/Nuclear-Power-Prevents-Deaths-Causes.html (2 April 2013).

42 Kavlak, G., McNerney, J. & Trancik, J. E. Evaluating the causes of cost reduction in photovoltaic modules. *Energy Policy* **123**, 700–710 (2018).

43 Kåberger, T. Progress of renewable electricity replacing fossil fuels. *Global Energy Interconnection* **1**, 48–52 (2018).

44 BP p.l.c., BP Energy Outlook 2017 Edition. (2017).

45 IEA. Global EV outlook 2018. https://www.iea.org/reports/global-ev-outlook-2018 Paris (IEA, 2018).

46 Newman, P. 1.1 – The renewable cities revolution. In *Urban Energy Transition* 2nd edn (ed Droege, P.) 11–30 (Elsevier, 2018). doi:10.1016/B978-0-08-102074-6.00015-2.

47 Hawkins, T. R., Singh, B., Majeau-Bettez, G. & Strømman, A. H. Comparative environmental life cycle assessment of conventional and electric vehicles. *Journal of Industrial Ecology* **17**, 53–64 (2013).

48 Sofiev, M. *et al.* Cleaner fuels for ships provide public health benefits with climate tradeoffs. *Nat Commun* **9**, 1–12 (2018).

49 Schäfer, A. W. *et al.* Technological, economic and environmental prospects of all-electric aircraft. *Nat. Energy* **4**, 160–166 (2019).

50 Lambert, F. A new all-electric and autonomous cargo ship is planned for operation in 2018. Electrek https://electrek.co/2017/05/11/all-electric-autonomous-cargo-ship/ (2017).

51 The world's first electric autonomous container ship to set sail in Norway. CleanTechnica https://cleantechnica.com/2018/08/23/the-worlds-first-electric-autonomous-container-ship-to-set-sail-in-norway/ (2018).

52 Lambert, F. Tesla Semi met and then crushed almost all of our expectations. Electrek https://electrek.co/2017/11/17/tesla-semi-electric-truck-specs-cost/ (2017).

53 Hao, K. Airbus, Rolls Royce, and Siemens are teaming up to build a hybrid-electric plane. Quartz https://qz.com/1139603/airbus-rolls-royce-and-siemens-are-partnering-to-create-a-hybrid-electric-plane/ (2017).

54 Gross, M. A planet with two billion cars. *Current Biology* **26**, R307–R310 (2016).

55 Berners-Lee, M. *How Bad Are Bananas? The Carbon Footprint of Everything* (Greystone Books, 2010).

56 IEA. World energy outlook, 2019. https://www.iea.org/reports/world-energy-outlook-2019 Paris (IEA, 2019).

57 Coady, D., Parry, I., Sears, L. & Shang, B. How large are global fossil fuel subsidies? *World Development* **91**, 11–27 (2017).

58 Erickson, P. *et al.* Why fossil fuel producer subsidies matter. *Nature* **578**, E1–E4 (2020).

59 Digiconomist. Bitcoin Energy Consumption Index. https://digiconomist .net/bitcoin-energy-consumption (accessed, 23 January 2019).

60 Vries, A. de. Bitcoin's growing energy problem. *Joule* **2**, 801–805 (2018).

61 MacKay, D. J. *Sustainable Energy – Without the Hot Air* (UIT Cambridge Ltd., 2009).

62 Musk, E. & Straube, J. B. Model S efficiency and range. Tesla https:// www.tesla.com/en_CA/blog/model-s-efficiency-and-range (2012).

63 Deutch, J. Decoupling economic growth and carbon emissions. *Joule* **1**, 3–5 (2017).

64 Gore, T. Extreme carbon inequality: why the Paris climate deal must put the poorest, lowest emitting and most vulnerable people first. *Oxfam International* https://www.oxfam.org/en/research/extreme-carbon -inequality. (2015).

65 Vaughan, A. Carbon emissions per person, by country. *The Guardian* http://www.theguardian.com/environment/datablog/2009/sep/02 /carbon-emissions-per-person-capita (2 September 2009).

66 O'Neill, B. C. *et al.* Global demographic trends and future carbon emissions. *PNAS* **107**, 17521–17526 (2010).

67 Kotler, S. The five-year ban: because a billion less people is a great place to start. *Psychology Today* http://www.psychologytoday.com/blog /the-playing-field/200902/the-five-year-ban-because-billion-less-people -is-great-place-start (8 February 2009).

68 Lutz, W., O'Neill, B. C. & Scherbov, S. Europe's population at a turning point. *Science* **299**, 1991–1992 (2003).

69 Hartmann, B. *Reproductive Rights and Wrongs. The Global Politics of Population Control* (South End Press, 1995).

70 Dyck, E. & Lux, M. Population control in the 'Global North'? Canada's response to Indigenous reproductive rights and neo-eugenics. *The Canadian Historical Review* **97**, 481–512 (2016).

71 Stephenson, J., Newman, K. & Mayhew, S. Population dynamics and climate change: what are the links? *J Public Health (Oxf)* **32**, 150–156 (2010).

72 Detraz, N. *Gender and the Environment* (Polity Press, 2016).

73 Bradshaw, C. J. A. & Brook, B. W. Reply to O'Neill et al. and O'Sullivan: fertility reduction will help, but only in the long term. *PNAS* **112**, E508–E509 (2015).

74 Bongaarts, J. & Sinding, S. W. A response to critics of family planning programs. *International Perspectives on Sexual and Reproductive Health* **35**, 039–044 (2009).

75 Muttarak, R. & Lutz, W. Is education a key to reducing vulnerability to natural disasters and hence unavoidable climate change? *Ecology and Society* **19** (2014).

76 Lutz, W., Muttarak, R. & Striessnig, E. Universal education is key to enhanced climate adaptation. *Science* **346**, 1061–1062 (2014).

77 IEA. World energy outlook, 2016. https://www.iea.org/reports/world -energy-outlook-2016 (Paris, 2016).

78 Smith, W. & Wagner, G. Stratospheric aerosol injection tactics and costs in the first 15 years of deployment. *Environ. Res. Lett.* **13**, 124001 (2018).

79 Intergovernmental Panel on Climate Change. *Global warming of 1.5°C.* (2018).

80 Keith, D. W., Wagner, G. & Zabel, C. L. Solar geoengineering reduces atmospheric carbon burden. *Nature Climate Change* **7**, 617–619 (2017).

81 Young, J. R. *et al.* A guide to extant coccolithophore taxonomy. *Journal of Nannoplankton Research, Special Issue* **1**, 1–132 (2003).

82 Zeebe, R. E. History of seawater carbonate chemistry, atmospheric CO_2, and ocean acidification. *Annu. Rev. Earth Planet. Sci.* **40**, 141–165 (2012).

83 Ridgwell, A. & Zeebe, R. E. The role of the global carbonate cycle in the regulation and evolution of the Earth system. *Earth and Planetary Science Letters* **234**, 299–315 (2005).

84 Doney, S. C., Fabry, V. J., Feely, R. A. & Kleypas, J. A. Ocean acidification: the other CO_2 problem. *Annual Review of Marine Science* **1**, 169–192 (2009).

85 Sperry, J. S. *et al.* The impact of rising CO_2 and acclimation on the response of US forests to global warming. *PNAS* **116**, 25734–25744 (2019).

86 Uddling, J., Broberg, M. C., Feng, Z. & Pleijel, H. Crop quality under rising atmospheric CO_2. *Current Opinion in Plant Biology* **45**, 262–267 (2018).

87 Giguère-Croteau, C. *et al.* North America's oldest boreal trees are more efficient water users due to increased $[CO_2]$, but do not grow faster. *PNAS* **116**, 2749–2754 (2019).

88 Balaraman, K. Whales keep carbon out of the atmosphere. *Scientific American* https://www.scientificamerican.com/article/whales-keep-carbon -out-of-the-atmosphere/ (11 April 2017).

89 International Monetary Fund. Nature's solution to climate change – IMF F&D. https://www.imf.org/external/pubs/ft/fandd/2019/12/natures -solution-to-climate-change-chami.htm.

90 Chatterjee, A. *et al.* Influence of El Niño on atmospheric CO_2 over the tropical Pacific Ocean: findings from NASA's OCO-2 mission. *Science* **358**, eaam5776 (2017).

91 Liu, J. *et al.* Contrasting carbon cycle responses of the tropical continents to the 2015–2016 El Niño. *Science* **358**, eaam5690 (2017).

92   Eldering, A. *et al.* The Orbiting Carbon Observatory-2 early science investigations of regional carbon dioxide fluxes. *Science* **358**, eaam5745 (2017).

93   Cox, P. M., Betts, R. A., Jones, C. D., Spall, S. A. & Totterdell, I. J. Acceleration of global warming due to carbon-cycle feedbacks in a coupled climate model. *Nature* **408**, 184–187 (2000).

94   Le Quéré, C. *et al.* Trends in the sources and sinks of carbon dioxide. *Nature Geoscience* **2**, 831–836 (2009).

95   Veen, C. J. van der. Fourier and the "greenhouse effect." *Polar Geography* (2008).

96   Tyndall, J. On the transmission of heat of different qualities through gases of different kinds. In *Notices of the Proceedings* Vol. 3 (William Clowes and Sons, 1859).

97   McNeill, L. This lady scientist defined the greenhouse effect but didn't get the credit, because sexism. *Smithsonian Magazine* https://www.smithsonianmag.com/science-nature/lady-scientist-helped-revolutionize-climate-science-didnt-get-credit-180961291/ (5 December 2016).

98   Jackson, R. Eunice Foote, John Tyndall and a question of priority. *Notes and Records: the Royal Society Journal of the History of Science* **74**, 105–118 (2020).

99   Molina, M. J. & Rowland, F. S. Stratospheric sink for chlorofluoromethanes: chlorine atom-catalysed destruction of ozone. *Nature* **249**, 810 (1974).

100 Farman, J. C., Gardiner, B. G. & Shanklin, J. D. Large losses of total ozone in Antarctica reveal seasonal ClO_x/NO_x interaction. *Nature* **315**, 207 (1985).

101 Schrope, M. Successes in fight to save ozone layer could close holes by 2050. *Nature* **408**, 627 (2000).

102 Chipperfield, M. P. *et al.* Detecting recovery of the stratospheric ozone layer. *Nature* **549**, 211–218 (2017).

103 Cheng, L., Abraham, J., Hausfather, Z. & Trenberth, K. E. How fast are the oceans warming? *Science* **363**, 128–129 (2019).

104 Myhre, G. *et al.* Anthropogenic and natural radiative forcing. In *Climate Change 2013. The Physical Science Basis. Contribution of Working Group I to the Fifth Assessment Report of the Intergovernmental Panel on Climate Change* (Cambridge University Press, 2013).

105 Williams, P. D. Increased light, moderate, and severe clear-air turbulence in response to climate change. *Adv. Atmos. Sci.* **34**, 576–586 (2017).

106 Williams, P. D. & Joshi, M. M. Intensification of winter transatlantic aviation turbulence in response to climate change. *Nature Climate Change* **3**, 644–648 (2013).

107 BBC News. Phoenix flights cancelled because it's too hot for planes. https://www.bbc.com/news/world-us-canada-40339730 (20 June 2017).

108 Thompson, A. German startup successfully tests electric VTOL "flying taxi." *Popular Mechanics* https://www.popularmechanics.com/flight /a26173/german-startup-tests-electric-vtol-taxi/ (2017).

109 Stern, P. C., Perkins, J. H., Sparks, R. E. & Knox, R. A. The challenge of climate-change neoskepticism. *Science* **353**, 653–654 (2016).

110 Marotzke, J. Quantifying the irreducible uncertainty in near-term climate projections. *Wiley Interdisciplinary Reviews: Climate Change* **10**, e563 (2019).

111 Lean, J. L. Observation-based detection and attribution of 21st century climate change. *Wiley Interdisciplinary Reviews: Climate Change* **9**, e511 (2018).

112 Maxton, G. & Randers, J. *Reinventing Prosperity. Managing Economic Growth to Reduce Unemployment, Inequality, and Climate Change* (Greystone Books, 2016).

113 Klein, N. *This Changes Everything* (Knopf Canada, 2014).

114 Truelove, H. B., Carrico, A. R., Weber, E. U., Raimi, K. T. & Vandenbergh, M. P. Positive and negative spillover of pro-environmental behavior: an integrative review and theoretical framework. *Global Environmental Change* **29**, 127–138 (2014).

115 Sanne, C. Willing consumers – or locked-in? Policies for a sustainable consumption. *Ecological Economics* **42**, 273–287 (2002).

116 Weber, C. L., Peters, G. P., Guan, D. & Hubacek, K. The contribution of Chinese exports to climate change. *Energy Policy* **36**, 3572–3577 (2008).

117 Helveston, J. & Nahm, J. China's key role in scaling low-carbon energy technologies. *Science* **366**, 794–796 (2019).

118 Wang, H. *et al.* China's CO_2 peak before 2030 implied from characteristics and growth of cities. *Nat. Sustain.* **2**, 748–754 (2019).

119 Nixon, R. *Slow Violence and the Environmentalism of the Poor* (Harvard University Press, 2013).

120 Postman, N. *Amusing Ourselves to Death. Public Discourse in the Age of Show Business* (Penguin Books, 1986).

121 Mauritsen, T. & Pincus, R. Committed warming inferred from observations. *Nature Climate Change* **7**, 652–655 (2017).

122 Raftery, A. E., Zimmer, A., Frierson, D. M. W., Startz, R. & Liu, P. Less than 2 °C warming by 2100 unlikely. *Nature Climate Change* **7**, 637–641 (2017).

123 Mortillaro, N. The psychology of climate change: why people deny the evidence. *CBC* https://www.cbc.ca/news/technology/climate -change-psychology-1.4920872 (2 December 2018).

124 Majumdar, A. & Deutch, J. Research opportunities for CO_2 utilization and negative emissions at the gigatonne scale. *Joule* **2**, 805–809 (2018).

125 Robin, L. Environmental humanities and climate change: understanding humans geologically and other life forms ethically. *Wiley Interdisciplinary Reviews: Climate Change* **9**, e499 (2018).

126 Ghosh, A. *The Great Derangement: Climate Change and the Unthinkable* (University of Chicago Press, 2016).

127 Haberkorn, T. Climate change: the coming calamity. *Die Zeit* (7 November 2018).

128 Gehl, L. *Claiming Anishinaabe. Decolonizing the Human Spirit* (University of Regina Press, 2017).

129 McCright, A. M. & Xiao, C. Gender and environmental concern: insights from recent work and for future research. *Society & Natural Resources* **27**, 1109–1113 (2014).

130 McCright, A. M. & Dunlap, R. E. The politicization of climate change and polarization in the American public's views of global warming, 2001–2010. *The Sociological Quarterly* **52**, 155–194 (2011).

131 Best, H. & Lanzendorf, M. Division of labour and gender differences in metropolitan car use: an empirical study in Cologne, Germany. *Journal of Transport Geography* **13**, 109–121 (2005).

132 Räty, R. & Carlsson-Kanyama, A. Energy consumption by gender in some European countries. *Energy Policy* **38**, 646–649 (2010).

133 Lamb, W. F. & Steinberger, J. K. Human well-being and climate change mitigation. *Wiley Interdisciplinary Reviews: Climate Change* **8**, e485 (2017).

134 Mazur, A. & Rosa, E. Energy and life-style. *Science* **186**, 607–610 (1974).

135 Mazur, A. Does increasing energy or electricity consumption improve quality of life in industrial nations? *Energy Policy* **39**, 2568–2572 (2011).

136 Morrison, T. H. *et al.* Mitigation and adaptation in polycentric systems: sources of power in the pursuit of collective goals. *Wiley Interdisciplinary Reviews: Climate Change* **8**, e479 (2017).

137 Struzik, E. *Firestorm. How Wildfire Will Shape Our Future* (Island Press, 2017).

138 Schiermeier, Q. Landmark court ruling tells Dutch government to do more on climate change. *Nature News* doi:10.1038/nature.2015.17841 (24 June 2015).

139 Khan, T. Climate change battles are increasingly being fought, and won, in court. *The Guardian* (8 March 2017).

140 Milman, O. New York City plans to divest $5bn from fossil fuels and sue oil companies. *The Guardian* (10 January 2018).

141 Sutterud, T. & Ulven, E. Norway sued over Arctic oil exploration plans. *The Guardian* (14 November 2017).

142 Robin, L. Environmental humanities and climate change: understanding humans geologically and other life forms ethically. *Wiley Interdisciplinary Reviews: Climate Change* **9**, e499 (2018).

143 Kendall, C. Ecuadorians to vote on constitution making its nature a rights-bearing entitiy. *The Guardian* (2008).

144 O'Donnell, E. & Talbot-Jones, J. Three rivers are now legally people – but that's just the start of looking after them. *The Conversation* http://theconversation.com/three-rivers-are-now-legally-people-but-thats-just-the-start-of-looking-after-them-74983 (23 March 2017).

145 Watts, J. Indigenous groups win greater climate recognition at Bonn summit. *The Guardian* (15 November 2017).

146 Eisenstein, M. How social scientists can help to shape climate policy. *Nature* **551**, S142 (2017).

147 Fankhauser, S. & Jotzo, F. Economic growth and development with low-carbon energy. *Wiley Interdisciplinary Reviews: Climate Change* **9**, e495 (2018).

148 Goldthau, A. The G20 must govern the shift to low-carbon energy. *Nature News* **546**, 203 (2017).

149 Polls reveal citizens' support for Energiewende. Clean Energy Wire. https://www.cleanenergywire.org/factsheets/polls-reveal-citizens-support-energiewende (accessed, 23 January 2019).

150 Leahy, S. Half of U.S. spending power behind Paris climate agreement. *National Geographic News* https://news.nationalgeographic.com/2017/11/were-still-in-paris-climate-agreement-coalition-bonn-cop23/ (15 November 2017).

151 Stern, P. C. *et al.* Opportunities and insights for reducing fossil fuel consumption by households and organizations. *Nature Energy* **1**, 16043 (2016).

152 Dietz, S., Bowen, A., Dixon, C. & Gradwell, P. "Climate value at risk" of global financial assets. *Nature Climate Change* **6**, 676–679 (2016).

153 Gambino, L. "Shortsighted, wrong": Apple, Facebook among tech giants to reject Paris pullout. *The Guardian* (2 June 2017).

154 Cherian, M. Scotland is now coal-free after shutting off its last coal power plant. Global Citizen https://www.globalcitizen.org/en/content/scotland-closes-longannet-last-coal-energy-plant/ (29 March 2016).

155 BBC. UK Parliament declares climate emergency. https://www.bbc.com/news/uk-politics-48126677 (1 May 2019).

156 The Guardian view on a Green New Deal: we need it now. Editorial. *The Guardian* (12 May 2019).

157 Ocasio-Cortez, A. Text: H.Res.109 – 116th Congress (2019–2020): Recognizing the duty of the Federal Government to create a Green New Deal.

https://www.congress.gov/bill/116th-congress/house-resolution/109/text (2019).

158 Penn, J. L., Deutsch, C., Payne, J. L. & Sperling, E. A. Temperature-dependent hypoxia explains biogeography and severity of end-Permian marine mass extinction. *Science* **362**, eaat1327 (2018).

159 Chakrabarty, D. Climate and capital: on conjoined histories. *Critical Inquiry* **41**, 1–23 (2014).

160 Satish, U. *et al.* Is CO_2 an indoor pollutant? Direct effects of low-to-moderate CO_2 concentrations on human decision-making performance. *Environ Health Perspect* **120**, 1671–1677 (2012).

161 Allen, J. G. *et al.* Associations of cognitive function scores with carbon dioxide, ventilation, and volatile organic compound exposures in office workers: a controlled exposure study of green and conventional office environments. *Environ Health Perspect* **124**, 805–812 (2016).

162 Jacobson, T. A. *et al.* Direct human health risks of increased atmospheric carbon dioxide. *Nat. Sustain.* **2**, 691–701 (2019).

163 Xiong, Y. *et al.* Indoor air quality in green buildings: a case-study in a residential high-rise building in the northeastern United States. *Journal of Environmental Science and Health, Part A* **50**, 225–242 (2015).

164 Lockwood, C. Building the green way. *Harvard Business Review* (June 2006).

165 Haines, A. & Ebi, K. The imperative for climate action to protect health. *New England Journal of Medicine* **380**, 263–273 (2019).

166 IEA. Tracking clean energy progress. https://www.iea.org/tcep/ (2017).

167 Zeng, N. Carbon sequestration via wood burial. *Carbon Balance and Management* **3**, 1 (2008).

168 Chiaramonti, D. Sustainable aviation fuels: the challenge of decarbonization. *Energy Procedia* **158**, 1202–1207 (2019).

169 Sterman, J. D., Siegel, L. & Rooney-Varga, J. N. Does replacing coal with wood lower CO_2 emissions? Dynamic lifecycle analysis of wood bioenergy. *Environ. Res. Lett.* **13**, 015007 (2018).

170 Raimi, D. Going deep on carbon capture, utilization, and storage (CCUS), with Julio Friedmann. (Resources Radio). https://www.resourcesmag.org/resources-radio/going-deep-carbon-capture-utilization-and-storage-ccus-julio-friedmann/ (accessed, 6 May 2020).

171 Markewitz, P. *et al.* Worldwide innovations in the development of carbon capture technologies and the utilization of CO_2. *Energy & Environmental Science* **5**, 7281–7305 (2012).

172 IEA. Carbon capture, utilisation and storage. https://www.iea.org/topics/carbon-capture-and-storage/capture/ (2019).

173   Greer, K. *et al.* Global trends in carbon dioxide (CO_2) emissions from fuel combustion in marine fisheries from 1950 to 2016. *Marine Policy* **107**, 103382 (2019).

174   McGrath, M. Climate change "magic bullet" gets boost. BBC News (3 April 2019).

175   National Academies of Sciences, Engineering, Medicine. *Negative Emissions Technologies and Reliable Sequestration. A Research Agenda.* doi:10.17226/25259 (2018).

176   Alcalde, J. *et al.* Estimating geological CO_2 storage security to deliver on climate mitigation. *Nature Communications* **9**, 2201 (2018).

177   Bellamy, R. & Geden, O. Govern CO_2 removal from the ground up. *Nat. Geosci.* **12**, 874–876 (2019).

178   CCUS Projects Network. EU CCS Demonstration Project Network. https://ccsnetwork.eu/ (accessed, 29 May 2019).

179   Roberts, D. Could squeezing more oil out of the ground help fight climate change? *Vox* https://www.vox.com/energy-and-environment/2019/10/2/20838646/climate-change-carbon-capture-enhanced-oil-recovery-eor (6 December 2019).

180   Núñez-López, V. & Moskal, E. Potential of CO_2-EOR for near-term decarbonization. *Front. Clim.* **1**, (2019).

181   Herzog, H. J. *Carbon Capture* (MIT Press, 2018).

182   Holden, E. Exxon sowed doubt about climate crisis, House Democrats hear in testimony. *The Guardian* (23 October 2019).

183   Parkin, B. Germany revives underground carbon capture plan in sign of climate struggle. *Bloomberg News* (20 May 2019).

184   Lim, X. How to make the most of carbon dioxide. *Nature News* **526**, 628 (2015).

185   Armstrong, K. & Styring, P. Assessing the potential of utilization and storage strategies for post-combustion CO_2 emissions reduction. *Front. Energy Res.* **3** (2015).

186   Zimmermann, A. & Kant, M. CO_2 utilisation today: report 2017. https://depositonce.tu-berlin.de/handle/11303/6247 (TU Berlin, 2017)

187   United Nations, Green economy. UN Sustainable Development Knowledge Platform https://sustainabledevelopment.un.org/index.php?menu=1446 (accessed, 30 January 2019).

188   Olah, G. A. *et al.* The continuing need for carbon fuels, hydrocarbons and their products. In *Beyond Oil and Gas. The Methanol Economy* 65–76 (John Wiley & Sons, Ltd, 2009). doi:10.1002/9783527627806.ch6.

189   Ozin, G. A case for CO_2-sourced sustainable synthetic fuels. *Advanced Science News* https://www.advancedsciencenews.com/a_case_for_co2_sourced_sustainable_synthetic_fuels/ (22 September 2017).

190 Grolms, M. Zero-emission shipping. *Advanced Science News* https://
www.advancedsciencenews.com/zero-emission-shipping/ (2 November
2018).

191 Jia, X., Klemeš, J. J., Varbanov, P. S. & Wan Alwi, S. R. Analyzing the energy consumption, GHG emission, and cost of seawater desalination in
China. *Energies* **12**, 463 (2019).

192 Caldera, U. & Breyer, C. Strengthening the global water supply through a
decarbonised global desalination sector and improved irrigation systems.
Energy **200**, 117507 (2020).

193 Carter, N. T. Desalination and membrane technologies: Federal research
and adoption issues. Congressional Research Service Report (2 January
2018).

194 Abdelkareem, M. A., El Haj Assad, M., Sayed, E. T. & Soudan, B. Recent
progress in the use of renewable energy sources to power water desalination plants. *Desalination* **435**, 97–113 (2018).

195 Zhang, L., Tang, B., Wu, J., Li, R. & Wang, P. Hydrophobic light-to-heat
conversion membranes with self-healing ability for interfacial solar heating. *Advanced Materials* **27**, 4889–4894 (2015).

196 Liu, Z. *et al.* Continuously producing watersteam and concentrated brine
from seawater by hanging photothermal fabrics under sunlight. *Advanced
Functional Materials* **29**, 1905485 (2019).

197 Izquierdo-Ruiz, F., Otero-de-la-Roza, A., Contreras-García, J., Prieto-
Ballesteros, O. & Recio, J. M. Effects of the CO_2 guest molecule on the *sI*
clathrate hydrate structure. *Materials* **9**, 777 (2016).

198 Yin, H. *et al.* Faradaically selective membrane for liquid metal displacement batteries. *Nature Energy* **3**, 127 (2018).

199 Kätelhön, A., Meys, R., Deutz, S., Suh, S. & Bardow, A. Climate change
mitigation potential of carbon capture and utilization in the chemical industry. *PNAS* **116**, 11187–11194 (2019).

200 Guminski, A., Böing, F., Murmann, A. & Roon, S. von. System effects of
high demand-side electrification rates: a scenario analysis for Germany
in 2030. *Wiley Interdisciplinary Reviews: Energy and Environment* **8**,
e327 (2019).

201 Qiao, Y. *et al.* Li-CO_2 electrochemistry: a new strategy for CO_2 fixation
and energy storage. *Joule* **1**, 359–370 (2017).

202 Olah, G. A. *et al.* The "methanol economy": general aspects. In *Beyond Oil
and Gas. The Methanol Economy* 179–184 (John Wiley & Sons, Ltd, 2009).

203 Wei, J. *et al.* Directly converting CO_2 into a gasoline fuel. *Nature Communications* **8**, 15174 (2017).

204 Fertilizer Canada. Advocacy. https://fertilizercanada.ca/advocacy/
(accessed, 24 January 2019).

205 Wang, L. *et al.* Greening ammonia toward the solar ammonia refinery. *Joule* **2**, 1055–1074 (2018).

206 Center for International Environmental Law. Plastic and Climate: The Hidden Cost of a Plastic Planet. http://www.ciel.org/plasticandclimate (accessed, 3 June 2019).

207 Tullo, A. Plastic has a problem; is chemical recycling the solution? *Chemical & Engineering News* http://cen.acs.org/environment/recycling /Plastic-problem-chemical-recycling-solution/97/i39 (6 October 2019).

208 Raman, S. K., Raja, R., Arnold, P. L., Davidson, M. G. & Williams, C. K. Waste not, want not: CO_2 (re)cycling into block polymers. *Chem. Commun.* **15**, 7315–7318 (2019) doi:10.1039/C9CC02459J.

209 Faruk, O. *et al. Lightweight and Sustainable Materials for Automotive Applications* (CRC Press, 2017).

210 Lehne, J. & Preston, F. Making concrete change: innovation in low-carbon cement and concrete. https://reader.chathamhouse.org/making -concrete-change-innovation-low-carbon-cement-and-concrete# (Chatham House, June 2018).

211 CO_2 Science and the Global CO_2 Initiative: global roadmap for implementing CO_2 utilization (7 December 2016).

212 Ellis, L. D., Badel, A. F., Chiang, M. L., Park, R. J.-Y. & Chiang, Y.-M. Toward electrochemical synthesis of cement: an electrolyzer-based process for decarbonating $CaCO_3$ while producing useful gas streams. *Proceedings of the National Academy of Sciences* **117**, 12584–12591 (2020).

213 UBC-led project combats emissions by locking carbon dioxide in mine waste. *UBC News* https://news.ubc.ca/2019/07/23/ubc-led-project- locks-carbon-dioxide-in-mine-waste/ (23 July 2019).

214 Rhodes, R. *Energy. A Human History* (Simon & Schuster, 2018).

215 FONA (Forschung für Nachhaltige Entwicklung). Carbon2Chem. https:// www.fona.de/en/carbon2chem-21629.html (accessed, 1 January 2019).

216 Collins, L. "World first" as hydrogen used to power commercial steel production. *Recharge.* https://www.rechargenews.com/transition /-world-first-as-hydrogen-used-to-power-commercial-steel-production /2-1-799308 (28 April 2020).

217 Bender, M., Roussiere, T., Schelling, H., Schuster, S. & Schwab, E. Coupled production of steel and chemicals. *Chemie Ingenieur Technik* **90**, 1782–1805 (2018).

218 Xu, M., Zhang, T. & Allenby, B. How much will China weigh? Perspectives from consumption structure and technology development. *Environ. Sci. Technol.* **42**, 4022–4028 (2008).

219 Chen, Q., Gu, Y., Tang, Z., Wei, W. & Sun, Y. Assessment of low-carbon iron and steel production with CO_2 recycling and utilization technologies: a case study in China. *Applied Energy* **220**, 192–207 (2018).

220 Licht, S. *et al.* Carbon nanotubes produced from ambient carbon dioxide for environmentally sustainable lithium-ion and sodium-ion battery anodes. *ACS Cent. Sci.* **2**, 162–168 (2016).

221 Wang, Q., Chen, X., Jha, A. N. & Rogers, H. Natural gas from shale formation: the evolution, evidences and challenges of shale gas revolution in United States. *Renewable and Sustainable Energy Reviews* **30**, 1–28 (2014).

222 The World Bank. Zero routine flaring by 2030. https://www.worldbank.org/en/programs/zero-routine-flaring-by-2030 (accessed, 1 January 2020).

223 Tavasoli, A. & Ozin, G. Green syngas by solar dry reforming. *Joule* **2**, 571–575 (2018).

224 Campbell, I., Kalanki, A. & Sachar, S. How to counter the climate threat from room air conditioners. www.rmi.org/insight/solving_the_global_cooling_challenge Rocky Mountain Institute (2018).

225 Dittmeyer, R., Klumpp, M., Kant, P. & Ozin, G. Crowd oil not crude oil. *Nature Communications* **10**, 1818 (2019).

226 Artz, J. *et al.* Sustainable conversion of carbon dioxide: an integrated review of catalysis and life cycle assessment. *Chem. Rev.* **118**, 434–504 (2018).

227 Carpenter, J. E. Peer-reviewed surveys indicate positive impact of commercialized GM crops. *Nature Biotechnology* **28**, 319–321 (2010).

228 Ferber, D. Risks and benefits: GM crops in the cross hairs. *Science* **286**, 1662–1666 (1999).

229 Arning, K. *et al.* Same or different? Insights on public perception and acceptance of carbon capture and storage or utilization in Germany. *Energy Policy* **125**, 235–249 (2019).

230 Pacala, S. & Socolow, R. Stabilization wedges: solving the climate problem for the next 50 years with current technologies. *Science* **305**, 968–972 (2004).

231 Peters, M. *et al.* Chemical technologies for exploiting and recycling carbon dioxide into the value chain. *ChemSusChem* **4**, 1216–1240 (2011).

232 Dinh, C.-T. *et al.* CO_2 electroreduction to ethylene via hydroxide-mediated copper catalysis at an abrupt interface. *Science* **360**, 783–787 (2018).

233 Xu, Y. *et al.* Selective production of aromatics from CO_2. *Catal. Sci. Technol.* **9**, 593–610 (2019).

234 Esrafilzadeh, D. *et al.* Room temperature CO_2 reduction to solid carbon species on liquid metals featuring atomically thin ceria interfaces. *Nat. Commun.* **10**, 1–8 (2019).

235 Dong, Y. *et al.* Low-temperature molten-salt production of silicon nanowires by the electrochemical reduction of $CaSiO_3$. *Angewandte Chemie International Edition* **56**, 14453–14457 (2017).

236 Cann, O. These are the top 10 emerging technologies of 2017. World Economic Forum https://www.weforum.org/agenda/2017/06/these-are-the-top-10-emerging-technologies-of-2017/ (26 June 2017).

237 Wynes, S. & Nicholas, K. A. The climate mitigation gap: education and government recommendations miss the most effective individual actions. *Environ. Res. Lett.* **12**, 074024 (2017).

238 Gooch, M. *et al*. The avoidable crisis of food waste: technical report. Value Chain Management International and Second Harvest (17 January 2019).

239 Ivanova, D. *et al*. Environmental impact assessment of household consumption. *Journal of Industrial Ecology* **20**, 526–536 (2016).

240 Hoicka, C. E. & Das, R. Ambitious deep energy retrofits of buildings to accelerate the 1.5 °C energy transition in Canada. *The Canadian Geographer*, 1–12 (2020) doi:10.1111/cag.12637.

241 Hischier, R. Car vs. packaging: a first, simple (environmental) sustainability assessment of our changing shopping behaviour. *Sustainability* **10**, 3061 (2018).

242 Weideli, D. *Environmental analysis of US online shopping*. MIT Center for Transportation & Logistics https://ctl.mit.edu/pub/thesis/environmental-analysis-us-online-shopping (2013).

Illustration and Text Credits

Figures were provided by Mireille Ghoussoub except for the following:

Figure 2: Keeling Curve, showing atmospheric carbon dioxide, in parts per million, from the years 1700 to 2020. Courtesy of Scripps Institution of Oceanography, https://scripps.ucsd.edu/programs/keelingcurve/.

Figure 3: Volume equivalent to one tonne of CO_2 in relation to the size of a London double-decker bus. Illustration courtesy of Dr. Chenxi Qian.

Figure 4: Global shares of primary energy. Illustration created by Geoffrey A. Ozin and Mireille Ghoussoub, using data taken from IEA World Energy Balances, 2019.

Figure 5: *Emiliania huxleyi* type A coccolithophore. Reproduced from Young, J. R. *et al.* A guide to extant coccolithophore taxonomy. *Journal of Nannoplankton Research*, special issue **1**, 1–132 (2003) with permission from Dr. Jeremy Young, University College London.

Figure 8: Solar irradiance spectrum showing the spectral distribution of radiation emitted by the sun before the radiation enters the earth's atmosphere (white line), and the spectrum upon the radiation's reaching the earth's surface (light-gray area). Illustration created by Geoffrey A. Ozin and Mireille Ghoussoub, using data from the ASTM G173-03 Reference Spectra, https://rredc.nrel.gov/solar//spectra/am1.5/.

Figure 10: Radiation spectrum of earth as seen from above the earth's atmosphere. Illustration created by Geoffrey A. Ozin and Mireille Ghoussoub, using data obtained from Hanel, R. A. *et al.* The Nimbus 4 infrared spectroscopy experiment: 1. Calibrated thermal emission spectra. *Journal of Geophysical Research (1896–1977)* **77**, 2629–2641 (1972).

Figure 17: Illustration of an all-in-one solar distillation system for producing fresh water from salt water. Illustration courtesy of Dr. Chenxi Qian.

The text contains material reproduced from the following articles:

Ozin, G. SF_6 worries: the most potent and persistent greenhouse gas. *Advanced Science News* https://www.advancedsciencenews.com/sf6-worries-the-most-potent-and-persistent-greenhouse-gas/ (2019). Copyright Wiley-VCH Verlag GmbH & Co. KGaA. Reproduced with permission.

Ozin, G. In flight and fight: the turbulence carbon dioxide causes. *Advanced Science News* https://www.advancedsciencenews.com/in-flight-and-fight-the-turbulence-carbon-dioxide-causes/ (2018). Copyright Wiley-VCH Verlag GmbH & Co. KGaA. Reproduced with permission.

Ozin, G. CO_2 on the brain and the brain on CO_2. *Advanced Science News* https://www.advancedsciencenews.com/co2-on-the-brain-and-the-brain-on-co2/ (2016). Copyright Wiley-VCH Verlag GmbH & Co. KGaA. Reproduced with permission.

Ozin, G. Photothermal desalination. *Advanced Science News* https://www.advancedsciencenews.com/photothermal-desalination (2015). Copyright Wiley-VCH Verlag GmbH & Co. KGaA. Reproduced with permission.

Index

About the Authors

GEOFFREY A. OZIN

Geoffrey Ozin is a distinguished university professor at the University of Toronto, and Government of Canada Research Chair in Materials Chemistry and Nanochemistry. He currently leads the Solar Fuels Team at the University of Toronto. He has held positions as honorary professor at the Royal Institution of Great Britain and University College London; external adviser for the London Centre for Nanotechnology; Alexander von Humboldt Senior Scientist at the Max Planck Institute for Surface and Colloid Science and the Center for Functional Nanostructures at the Karlsruhe Institute of Technology; and Global Chair at Bath University. He is the author of two books: *Nanochemistry: A Chemical Approach to Nanomaterials* (2006) and *Concepts of Nanochemistry* (2009). He lives with his wife in Toronto, Canada.

MIREILLE F. GHOUSSOUB

Mireille Ghoussoub is a doctoral candidate in Materials Chemistry working with the Solar Fuels Team at the University of Toronto. Her research is focused on the study of CO_2 reaction pathways occurring on the surface of nanocrystalline catalysts, using computational and spectroscopic techniques. She completed her BASc in Engineering Physics at the University of British Columbia in Vancouver, and her MASc in Materials Science and Engineering at the University of Toronto. She lives in Toronto, Canada.